CHROMIUM PICOLINATE

EVERYTHING YOU NEED TO KNOW

CHROMIUM PICOLINATE

EVERYTHING YOU NEED TO KNOW

DR. GARY EVANS

Avery Publishing Group

Garden City Park, New York

Cover Design: Rudy Shur and William Gonzalez
In-House Editor: Lisa James
Typesetter: Bonnie Freid
Printer: Paragon Press, Honesdale, PA

Library of Congress Cataloging-in-Publication Data

Evans, Gary W. (Gary William), 1948-
 Chromium picolinate : everything you need to know / by Gary W.
Evans.
  p. cm.
 Includes bibliographical references and index.
 ISBN 0-89529-731-0
 1. Chromium in human nutrition. 2. Chromium—Physiological
effect. I. Title.
QP535.C7E93 1996
612.3'924—dc20   96-4107
                                                    CIP

Printed in the United States of America

10 9 8 7

CONTENTS

To all the researchers, students, and colleagues
whose work made this book possible

INTRODUCTION

Chromium just might be one of the most important nutrients in the body, but many people have never heard of it.

Body cells need chromium to keep insulin working properly. Insulin is a hormone—a substance that controls the body's chemical processes—secreted into the bloodstream after we eat. It directs the movement of digested food into the body's cells and affects how that food is used. When insulin doesn't act exactly as intended, blood sugar and fat aren't stored and used the way Nature planned. This can lead to obesity, heart disease, or diabetes.

But there's a catch. Our diets contain little chromium, and what chromium we do eat is often in a form that is difficult for the body to absorb. This means that most people don't get all the chromium they need to stay healthy. So chromium must be combined with a substance that will allow this insulin-regulating metal to easily enter the bloodstream. That substance is picolinate. Therefore, chromium picolinate is a form of chromium the body can use effectively.

Studies with both humans and animals have proven that chromium picolinate improves insulin's effectiveness. During a large study conducted in Israel, doctors found that insulin action was improved in diabetics after only ten days of chromium picolinate supplements.[1] In another study, insulin action was greatly improved when women with gestational diabetes—the kind that sometimes develops during pregnancy—were given chromium picolinate.[2] Studies with animals also showed these effects.

Dozens of other studies have proven that chromium, in the form of chromium picolinate, helps control blood fat, blood sugar, body fat, food cravings, and an age-related bone weakening called osteoporosis. It also stimulates muscle development and prolongs life. As wide-ranging as these effects seem, they all stem from chromium's ability to help insulin do its job.

What can chromium picolinate do? You might have heard of LDL and HDL cholesterol. LDL is the form of cholesterol that accumulates in the blood vessels and causes heart disease. HDL doesn't accumulate in blood vessels, so we want a large portion of our blood cholesterol to be in HDL form. One study showed that use of chromium picolinate results in lower levels of both total and LDL cholesterol, and in higher levels of HDL cholesterol.[3] Another study, one done with diabetics, showed that levels of both blood sugar and an abnormal blood chemical found in diabetics are lower when the patients take a daily chromium picolinate supplement.[4] These results have subsequently been reproduced by other doctors in independent studies.

There's more good news. Studies at three different universities showed that regular exercise coupled with chromium picolinate supplements produce a dramatic increase in muscle along with a marked reduction in body fat.[5–7] Another study with nonexercising individuals led to a pronounced loss of body fat with no loss of muscle after seventy-two days of chromium picolinate supplements.[8] And in an Austrian study, chromium picolinate prevented the muscle loss that often accompanies severe fat-reduction diets.[9]

Chromium picolinate supplements lead to greater production of anabolic steroids—the chemicals that allow our muscles to grow—in both men and women. Middle-aged men and postmenopausal women had markedly increased levels of a steroid hormone called DHEA while taking chromium picolinate.[10,11] Also, postmenopausal women who take chromium picolinate excrete much less calcium, which means that they are less likely to develop osteoporosis. As in the case of the diabetes studies, the results of all these human studies have been supported by animal studies.

Chromium was first recognized as an important nutrient over thirty years ago. But this knowledge could not be acted upon because scientists at that time did not know which form of chromium the body could use. I became interested in this subject about twenty-five years ago while working for the United States Department of Agriculture. I continued my research after leaving government service to become a college professor. These studies showed that picolinate was the key to unlocking chromium's potential (see Chapter 1).

In 1989, I made my findings public at a scientific meeting. Since that time, a great deal of sound scientific data has been accumulated (see Chapters 2 and 3). There has also been a great deal of misinformation generated, mostly by people who have never taken a look at the dozens of studies that show the benefits of chromium taken in the form of chromium picolinate.

This book is intended to gather together and explain the meaning of the studies that have been done with chromium picolinate, and to tell you how you can use this important nutrient to improve your own life. The first three chapters present study results. Chapter 4 gives a detailed explanation of why chromium picolinate helps you stay healthy, while Chapter 5 explains exactly how chromium helps insulin do its job. Chapters 6, 7, 8, and 9 show how you can use chromium picolinate to deal with different health concerns, and Chapter 10 contains some closing remarks.

I believe that all people can benefit from the use of chromium picolinate supplements to fight obesity, heart disease, and the general effects of aging. Women can especially benefit, as they run a greater risk of developing osteoporosis than men. Athletes can use chromium picolinate to safely increase muscle mass. Diabetics can use chromium picolinate to reduce the need for insulin and drugs, but *must* consult with their doctors first (see Chapter 7).

I hope this book will clear up misinformation about chromium picolinate, and that after reading it you will regard chromium as a forgotten nutrient that is finally gaining the recognition it deserves. I also hope this book helps you in your quest for good health.

1. WHAT IS CHROMIUM PICOLINATE?

I first learned about the vital link between chromium and health while working for the United States Department of Agriculture (USDA), and about a unique chemical called picolinate (pronounced *pick-coal-in-ate*). Picolinate is made from a protein component found in living cells, and it greatly increases the body's absorption and use of minerals such as zinc, copper, iron, and chromium.

Most of my USDA studies were with zinc picolinate (see Chapter 4). But I also worked with Dr. Walter Mertz, one of the pioneers in chromium research, who helped me understand the body's need for chromium. When I left government service to become a chemistry professor in 1982, I began working with chromium picolinate out of a belief that a lack of chromium in our diets is a leading cause of health problems in the world.

In this chapter, after some general information on chromium picolinate, we will discuss the human studies that I've done in the areas of reducing cholesterol, controlling diabetes, shaping the body, and strengthening bone, and about the existing research that I read about before doing my own studies. Studies done to follow up on my work are covered in Chapters 2 and 3.

CHROMIUM PICOLINATE: A COMBINATION OF TWO VERY IMPORTANT CHEMICALS

Chromium picolinate is a combination of the mineral chromium and a little-known chemical called picolinate. Many studies with chromium show it is needed to make the hormone insulin work efficiently.[1-6] Insulin, which regulates blood fat and sugar, can do its job without chromium. But when chromium exists in ample quantities, insulin works much, much more effectively.

However, a problem exists because there isn't enough chromium in the foods we eat.[7,8] What chromium does exist in food is often not absorbed and used by the body. The same can be said of food supplements or other products that have added chromium. The mere presence of chromium in food— or of any mineral, for that matter—is no guarantee that health benefits can be derived from it.

Why can't we always use the minerals we eat? Minerals exist inside the body in the form of ions, and chromium is a supercharged ion. It has to be made electrically neutral before it can approach and eventually enter living cells. The membrane "fence" that surrounds a cell repels anything that is charged, so ions have to have their charges neutralized.[9-12]

Chromium can be made acceptable to living cells when it is attached to special chemical substances, called chelators, that neutralize ions. Chelators play an important role in the delivery of minerals to places in the body where they're needed. While working for the USDA, my colleagues and I discovered that picolinate is a superior chelator.[13-29] Later, at Bemidji State University (BSU) in Minnesota, my students and I learned that picolinate very effectively cloaks the chromium ion's charge so that it can be smuggled into the bloodstream.

The Science of Studies

Medical tests are performed according to a strict set of procedures known as a protocol. This ensures that the test results will be as accurate as possible and that the tests can be repeated by other doctors.

Many tests measure the drug or other test item against a control. This gives the doctors conducting the test something to measure the drug's effectiveness against. In human tests, some of the participants are given the drug, while others are given an inert substance called a placebo. In animal studies, some of the animals are given the drug, while the rest are given nothing.

Sometimes, the people being tested are used as their own controls by first giving them either the drug or the placebo for a given period, taking a break, and then switching to the other substance. This is known as cross-over. A study is a double-blind cross-over when all the pills are in coded bottles and neither the doctors nor the participants know what the participants are taking until the end of the study, when the code is broken. This prevents false positive results caused by psychological effects, and guards against special treatment of either the volunteers or the data by the investigators.

This makes chromium picolinate a biologically available form of chromium—one that can be used by the body.

CHROMIUM PICOLINATE LOWERS CHOLESTEROL

Heart disease, a leading cause of death throughout the world, was one of the first medical problems to be associated with a lack of chromium in the body. In the late 1960s, Dr. Henry Schroeder of Vermont, a researcher associated with Dartmouth Medical School, established a link between chromium deficiency and atherosclerosis, or the buildup of cholesterol on blood vessel walls.[30–32] Dr. Schroeder found very little, if any, chromium in the aorta—the body's main artery—of people who died of heart disease. But there was chromium in the aortas of people who lived to the same age or longer and died of accidental causes.

In his book *The Poisons Around Us*, first published in 1974, Dr. Schroeder devoted the last chapter to vitamin and mineral deficiencies in this country.[33] In his discussion of chromium, Dr. Schroeder says that in several studies, cholesterol levels were lowered in about half the individuals treated with chromium salts. He goes on to explain how chromium salts are poorly absorbed, but that a natural form found in yeast is effective. He says, "Serious efforts are being made to purify the substance and make it: when that is done the day of specific treatment of atherosclerosis will dawn."

High cholesterol levels can lead to heart disease because one type of cholesterol—called LDL, or bad, cholesterol—sticks to the insides of blood vessels. This narrows the vessels, and sometimes a blood clot can become stuck at one of these narrow points, which can cause a heart attack. On the other hand, good HDL cholesterol can remove fat from the bloodstream.[34–36]

I read Dr. Schroeder's discussion of chromium and cholesterol with interest. Because cholesterol is directly related to the development of atherosclerosis and heart disease, a good way to test chromium picolinate's effectiveness in humans would be to test its effect on blood cholesterol. In 1988, Dr. Sheldon Hendler helped me arrange a collaboration with Dr. Ray Press and Dr. Jack Geller at Mercy Hospital in San Diego, California. The doctors at Mercy agreed to conduct two separate studies with chromium picolinate, one of which involved people with high blood-cholesterol levels.[37,38]

The Mercy study involved 28 volunteers who had abnormally high blood cholesterol levels. They were given either a placebo or 200 micrograms of chromium as picolinate daily for forty-two days, followed by a fourteen-day period with no supplement. Each of the volunteers then switched supplements for another forty-two-day period. (See "The Science of Studies" on page 7 for a description of how medical studies are conducted.)

Both total cholesterol and LDL cholesterol decreased when the participants were taking chromium picolinate. Total cholesterol decreased by 7 percent, while LDL cholesterol decreased by 10.5 percent. Of the 28 participants, 22 had lower total cholesterol levels, 2 had the same cholesterol levels, and the remaining 4 had slightly elevated cholesterol levels. Twenty individuals had decreased amounts of LDL cholesterol, 3 had unchanged levels, and 5 had slightly elevated levels. The amount of HDL cholesterol increased slightly for everyone. Absolutely nothing happened to cholesterol levels when the participants were taking the placebo.

We also noticed increased production of the protein that forms HDL cholesterol (see Chapter 9). Many cardiologists have told me that this increase is more meaningful in health terms than the drop in LDL cholesterol. After these studies, I was convinced we had discovered the form of chromium Dr. Schroeder had said would help eliminate heart disease.

DIABETICS BENEFIT
FROM CHROMIUM PICOLINATE

By 1960, chromium had been determined to be essential in animals because of its effect on blood sugar, the simple sugar that fuels the body's cells (see Chapter 5). But chromium's importance to glucose regulation in humans wasn't discovered until 1977, when diabetic symptoms in a woman being fed intravenously were eliminated by adding chromium to the IV fluid.[39]

Diabetes—the body's inability to control blood-glucose levels—takes several different forms (see Chapter 7). In one form, the body cannot use insulin, the hormone that controls blood sugar, very effectively. When insulin cannot be used effectively, glucose accumulates in the blood until it reaches very high levels, causing a condition known as hyperglycemia, or high blood sugar. These high levels of glucose cause many problems.

One problem that is currently being studied occurs when glucose attaches itself to various proteins in the body. One of these proteins is the oxygen-carrying protein hemoglobin. When glucose attaches to hemoglobin, the protein can be easily detected in the blood. This new protein is glycated hemoglobin—hemoglobin with glucose attached—and is widely used to detect changes in insulin action.

No test of chromium picolinate would have been complete without trials involving people with diabetes. So the doctors at Mercy Hospital who did the cholesterol tests did tests on diabetic patients.[40] Eleven volunteers were given either chromium picolinate or a placebo for six weeks.

Eight of the participants responded positively while they were using the chromium supplements. In the people who showed a positive response, blood-glucose levels decreased by 24 percent and glycated-hemoglobin levels decreased by 19 percent. These people also showed cholesterol-level decreases. Total cholesterol levels decreased by 13 percent and LDL cholesterol levels decreased by 11 percent.

The two studies at Mercy Hospital showed that chromium picolinate is more beneficial than other forms of chromium because a high percentage of the participants responded during a relatively short period of time. In most trials with other forms of chromium, less than half the participants respond to the supplement. But during the cholesterol trial with chromium picolinate, nearly 80 percent of the volunteers had lower total cholesterol levels after only six weeks, while a positive response was noted in 72 percent of the diabetics.

During one small but very meaningful study, 100 percent of the volunteers responded to chromium picolinate supplements. This study involved Native Americans living in the area of Minnesota where BSU is located. Native Americans have a high incidence of diabetes, and I wanted to see if chromium could help reduce this incidence. Unfortunately, it was difficult to set up a formal controlled study through the local Native American health-care network. After I mentioned this on a couple of occasions to my nutrition classes, one of my students decided to conduct a small study that didn't involve going through established channels.[41]

The student enlisted five members of the local Chippewa band, one man and four women. The participants took one capsule of chromium picolinate each day, and had their blood analyzed before they started the supplements and again every two weeks, for a total of eight weeks.

Blood-glucose levels were down considerably in each of the five participants after only two weeks. After eight weeks, blood-glucose levels decreased by 32.6 percent. Blood glucose decreased by 61 percent in one woman who was being treated with a large amount of insulin—75 Units a day. She was able to cut her insulin dosage. The cholesterol levels in these volunteers were not abnormally high, but still decreased by 8.2 percent. With their high rates of heart disease and diabetes, Native Americans could derive many health benefits from chromium picolinate supplements.

CHROMIUM PICOLINATE RESULTS IN BIGGER MUSCLES AND LESS FAT

The studies with people who had high cholesterol levels or diabetes proved that chromium picolinate is a very effective way for human beings to get enough useful chromium. The results were both revealing and exciting. But just as revealing were tests done with healthy, young athletes.

After the success with the studies at Mercy Hospital, we decided to test the effect of chromium picolinate supplements on muscle growth and development.[42] Insulin is sometimes referred to as an anabolic, or growth-stimulating, hormone, because it causes amino acids—protein's building blocks—and glucose to move into muscle cells (see Chapter 8). For the muscle studies, I was fortunate to have the assistance of Dr. Muriel Gilman, an exercise physiologist at BSU, and Guy Otte, a graduate student who was working with the university's football players.

In our first trial, ten male students were recruited from the school's weight-training classes. The young men were given, at random, a supply of either placebo or chromium picolinate and instructed to take one capsule a day for forty days. At the start and again at the end of the trial, height and weight were measured, along with biceps and calf circumferences. Also, the thickness of the skin—called the skin fold—was measured on various parts of the body to calculate the amount of muscle in the body, which physiologists call the lean body mass. During the course of the trial, the students maintained their normal eating habits and activity levels, while each participated in a weightlifting program of two forty-minute periods a week.

In this small preliminary study, the circumference of both the biceps and the calf increased more in the students who were given chromium picolinate than in those who took the placebo. In the chromium group, biceps circumference increased by 0.55 inches, compared with an increase of 0.47 inches for the placebo group. The calf circumference of the students who took chromium picolinate increased by 0.45 inches, compared with a placebo-group increase of only 0.32 inches.

There was also a gain in weight by both groups, a sign that students in both groups were building some muscle. But the weight gain was greater in the chromium picolinate group— an increase of about 4.75 pounds, compared with a gain of about 2.5 pounds for the placebo group. The increase in lean body mass of the students taking the chromium was about 3.5 pounds, but the increase in the students taking the placebo was less than 2 ounces! This shows that nearly three-fourths of the weight gain in the students given chromium picolinate was caused by an increase in muscle.

Although the participants did not know which substance they were taking, most of the young men taking the chromium supplement seemed to sense it. Before they were told who was getting what, the men taking chromium remarked that they noticed an increase in energy and a definite improvement in muscle development.

This study was very revealing in more ways than one. We found that hydrostatic weighing, or weighing a person in a special underwater chair, is not as accurate as many scientists think. Hydrostatic weighing is considered a reliable measure of muscle mass because fat floats but muscle doesn't. However, if the tests are not performed with great care, the results are useless. This happened in our study. Fortunately, Dr. Gilman had the foresight to take skin-fold measurements with skin calipers, so we were able to calculate the muscle-mass gains of the young men.

The first study with athletes was encouraging, but small. So I was happy when Otte asked if he could measure the effect of chromium picolinate on muscle development in football players during their offseason training period. This study involved 31 athletes: 16 taking chromium picolinate daily and the others taking the placebo. Everybody lifted weights for an hour and a half, five days a week for six weeks.

After only fourteen days, the group given the chromium had lost 2.7 percent of their body fat and gained 4 pounds of muscle. At the completion of the study, the athletes taking chromium had lost 7.5 pounds of fat, a total decrease of 23 percent. After six weeks of exercise and chromium supplements, muscle mass had increased by 5.7 pounds. The athletes taking the placebo lost 2.2 pounds of fat, but only gained 4 pounds of muscle.

While the trials with diabetes and cholesterol were very encouraging, the muscle studies conducted at BSU were a true discovery. Never before had any study shown that accelerated muscle development and fat loss could be achieved with a chromium supplement. What was especially encouraging is that the fat loss was achieved without any attempt to control the fat content of the participants' diets.

CHROMIUM PICOLINATE AND BRITTLE BONES

Over ten million women are afflicted with the bone problem called osteoporosis. The bones lose their density because calcium is lost through the urine. As a result, they become brittle and crack easily. While a lack of calcium is most often associated with osteoporosis, my students and I conducted an experiment that left us with the impression that chromium also plays a part in controlling this condition.

Osteoporosis occurs in women after menopause due in large part to the decreased production of female hormones. The most abundant steroid hormone—a hormone that controls many different bodily functions—in the bloodstream is called dehydroepiandrosterone (DHEA). In women, DHEA can be converted to estrogen. Without enough estrogen, calcium simply gets filtered through the kidneys and is washed out of the body. (For a more detailed discussion of this process, see Chapter 9.)

Studies show that postmenopausal women have high levels of insulin but much lower levels of DHEA than women twenty years younger, because insulin interferes with DHEA production.[43] Because of the connection between insulin and DHEA, we decided to analyze DHEA and estrogen levels in 27 women who had reached menopause.

At the beginning of the study, we measured levels of calcium in the urine along with levels of DHEA and estrogen in the blood. The women then started taking 400 micrograms of chromium in picolinate form daily. After a month, we again measured levels of calcium, DHEA, and estrogen.

Before the women started taking chromium picolinate, the DHEA and estrogen levels were comparable to those found in most women between the ages of forty and sixty. Urinary calcium loss was about twice what is found in women who have not reached menopause.

After taking chromium supplements for two months, DHEA and estrogen levels in the blood increased to levels more like those found in women between the ages of thirty and thirty-five. The amount of calcium excreted decreased by half. A month after the women stopped taking chromium picolinate, DHEA and estrogen levels decreased while urinary calcium excretion doubled.[44] This study provides strong evidence that chromium supplements can be used to prevent loss of calcium after menopause.

I was gratified to see evidence that chromium picolinate could help people lead healthier lives in so many ways. I was even more gratified when other researchers started to confirm these results.

2. CHROMIUM PICOLINATE IS FOR REAL— HUMAN STUDIES

High cholesterol levels and problems with blood-sugar regulation are pressing medical issues in this country. So our studies with chromium picolinate rekindled an interest in chromium among some health professionals. But many others, already aware of chromium's role in treating and preventing these conditions, felt that our studies provided no new information.

I expected this response, because most professionals didn't realize how effective chromium picolinate is compared with other forms of chromium. Our studies showed a positive response in nearly 75 percent of the participants, a response rate achieved by no other form of chromium. Most ignored the issue of whether or not the body can effectively use a mineral in a given form—a concept that scientists call bioavailability.

There was a stronger response among scientists to the effect of chromium picolinate on muscle development and fat loss. This discovery could benefit nearly 33 percent of the United States population—the percentage of people in this country who are overfat.

The term "overweight" is misleading. Muscle is heavier than fat, but fat is what causes obesity and the diseases that result from obesity. Thus, a person can be overweight according to the standard height-and-weight charts, but have a low percentage of body fat.

The potential use of chromium picolinate by athletes for building muscle also sparked interest, as this could eliminate the temptation for some athletes to use harmful synthetic anabolic steroids. I talked to concerned parents of teenaged boys who wanted to know if chromium picolinate could actually replace steroids. I talked with professional and amateur athletes. Everybody wanted to know how chromium picolinate works and if it was safe. Many also asked, "Are you going to repeat your trials?"

Our early studies involved one to two dozen people each, but studies with ten or twelve dozen people in each study would be more convincing. I realized that scientists, being scientists, would want to see if our studies could be repeated by others, and so decided to wait and see if somebody would repeat our work.

It wasn't a long wait. Reports of studies with humans and animals began appearing soon after our initial announcements. The number of studies steadily increased, and continues to do so. In this chapter, we'll look at results from human studies done in the areas of cholesterol reduction, muscle building and fat loss, and blood-sugar reduction in diabetics. In Chapter 3, we'll turn to the animal studies.

CLEANING THE BLOODSTREAM EASILY AND INEXPENSIVELY

One of the first of these other studies came from a doctor who discovered that chromium picolinate could be combined with niacin, also known as vitamin B3, to inexpensively and effectively control cholesterol. Dr. J. B. Gordon of San Diego, California, treated ten patients with a combination of chromium picolinate (200 micrograms) and one to two grams of niacin every day for a period of four weeks.[1]

The published results didn't say if all the patients in this study responded to treatment. But the results did say that total cholesterol levels dropped by 24 percent, LDL cholesterol levels—the bad cholesterol described in Chapter 9— dropped by 27 percent, and blood-fat levels dropped by 43 percent. Levels of HDL cholesterol—the good cholesterol— didn't change, but the ratio of LDL to HDL improved greatly, with HDL cholesterol representing a much higher percentage of total cholesterol. This is exactly what doctors want to see.

This trial was particularly meaningful because it usually requires more than three grams of niacin a day to get the kind of response noted by Dr. Gordon. Also, niacin often causes HDL cholesterol levels to decrease. That did not occur during the trial with chromium picolinate.

I believe there's a good explanation for the effectiveness of the chromium picolinate-niacin combination. Both picolinate and niacin are made from tryptophan, one of the protein building blocks known as amino acids.[2] Thus, the niacin supplement will divert all of the tryptophan present in the diet toward picolinate production. I think niacin is effective in reducing blood fats because the increased picolinate can be used to increase the effectiveness of what little chromium exists in the diet.

Dr. Nancy Lee and Dr. Charles Reasner at the University of Texas Health Science Center in San Antonio, Texas, examined the effect of chromium picolinate supplements containing 200 micrograms of chromium in a double-blind crossover study.[3] (See "The Science of Studies" on page 7 for an explanation of how medical studies are conducted.) The participants in this study were 28 diabetics.

While the published report again did not say if all the participants responded to the chromium picolinate, it did say that blood-fat levels decreased by more than 17 percent during the period the participants were given chromium picolinate. However, for some unknown reason, there was no improvement in either blood sugar or glycated hemoglobin, a blood chemical found in diabetics, during this particular study. As we'll see in the next section, I started to think at this point that 200 micrograms of chromium a day might not be enough to produce consistent results.

BUILDING MUSCLE AND LOSING FAT

Many of the human studies involving chromium picolinate have explored its usefulness in building muscle and eliminating fat. While these two aspects are related, the researchers who conducted these studies tended to concentrate on one or the other.

Chromium Picolinate and Muscle Gain

One study was conducted by a researcher with a personal interest in the subject. Deborah Hasten, a bodybuilder working on her doctoral degree at Louisiana State University (LSU), studied 59 college students in a weightlifting class.[4] The 22 women and 37 men were enrolled in a twelve-week program, twice as long as in our study.[5] In Hasten's double-blind study, half of the students were given chromium picolinate (200 micrograms of chromium) while the other half were given a placebo. Measurements were taken at the beginning and at the end of the trial.

In this study, the women given chromium picolinate gained, on average, five pounds of muscle, much more than either the men or the women who were taking the placebo. The sizes of the chest, arm, and thigh also increased more on these women.

The results for the men were a little different than what we had seen in our study. The sizes of the chest, arm, and thigh increased most on the men given chromium picolinate, but the increase in the total amount of muscle was the same in both groups—the men receiving chromium picolinate and those receiving the placebo—at the end of the study. I decided that the men didn't get enough chromium to make enough of a difference.

A number of athletes, particularly bodybuilders, confirmed this. They had told me that 200 micrograms of chromium was not enough to be effective with athletes who already had developed muscle mass. Most had started using 400 to 800 micrograms a day and were seeing better results. This explained why the women in our study showed a more dramatic increase—they generally had less developed muscle mass to begin with.

We had seen good results in our male athletes at the 200-microgram level, but our study lasted six weeks while the LSU study lasted twelve weeks. Keep in mind that in both studies, athletes taking chromium were compared with athletes taking a placebo, and both groups exercised. You expect exercise to increase muscle, and it did in both studies. I think the young men taking the placebo in the LSU study eventually caught up with the men taking the chromium because the amount of chromium wasn't high enough to achieve results after a certain period of time.

Deborah Hasten's studies were illuminating because they drew attention to the fact that you have to consider existing muscle mass before you can determine the amount of chromium needed. Animal studies—discussed in Chapter 3— raise the possibility that humans may require as much as 500 micrograms a day. (See Chapter 8 for a more in-depth look at chromium picolinate and sports.)

The Colgan Institute of Nutritional Science in San Diego is a consultation, education, and research facility concerned with effects of nutrition on athletic performance and aging. Dr. Michael Colgan, the institute's director, tested chromium picolinate in hundreds of athletes, both men and women, in trials that ran between six and twelve months.[6]

Colgan found the same effect in men reported by Deborah Hasten. When the men were given 200 micrograms of chromium a day, there was no improvement. However, when body weight was taken into consideration, and between 200 and 800 micrograms of chromium were used, Dr. Colgan saw reliable and consistent increases in muscle growth and strength, along with reductions in body fat. Dr. Colgan has found that only chromium in the form of chromium picolinate is effective, and no longer uses other forms of the mineral.

Chromium Picolinate Helps Promote Fat Loss Without Exercise

Our studies had shown that chromium picolinate could reduce the amount of body fat among bodybuilders and other athletes. But what about people who don't exercise?

That was the focus of a study conducted by Dr. Gil Kaats and a team of doctors at the Health and Medical Research Foundation and the University of Texas Health Science Center, both in San Antonio.[7] This was a double-blind study in which the effect of two different levels of chromium picolinate was studied by adding either the chromium or a placebo to a liquid drink. The 154 participants were told not to change their diets or exercise levels, or change the amount of food they ate. They were asked to take two of the special drinks a day.

Some of the volunteers who took the chromium picolinate received a low dose—an average of 200 micrograms of chromium a day. Others received a high dose—an average of 400 micrograms a day.

After seventy-two days, the group given the placebo showed no changes, but the changes in the chromium groups were astonishing. The 200-microgram group lost an average of 3.4 pounds body fat. But the 400-microgram group lost an average of 4.6 pounds body fat—about 35 percent more. The average loss of body fat for all those taking chromium picolinate was 4.2 pounds with an accompanying increase in lean body mass of 1.4 pounds. The average fat loss of the people who had the placebo in their drinks was only 0.8 pounds! As expected, those getting the higher level of chromium picolinate had the most dramatic changes in body composition, and older people (average age fifty-five) showed better improvements than younger people (average age thirty-six). This trial was very informative because it showed that chromium picolinate supplements help accelerate fat loss without exercise.

Dr. Kaats and his colleagues then conducted a second trial, which was equally informative. In the second study, 30 obese women and 10 obese men were recruited for a trial in which each participant served as his or her own control.[8] The experiment was divided into two phases, each lasting eight weeks. During the first phase, the women were put on a low-fat diet of 1,250 calories a day. The men were put on a diet of 1,650 calories. (See "What Is a Calorie?" on page 24 for a short explanation of what calories are.)

During this first phase, there were no notable changes in body composition or blood cholesterol. Metabolic rate—the amount of oxygen used by the body—was measured in 29 of the participants. In 23 of the 29, metabolic rate decreased slightly during this phase. This indicates that the body's cells weren't burning much fat.

In the second phase of this trial, the women and men were kept on the same low-calorie diets, but the diets were supplemented every day with two high-fiber cookies and two capsules containing chromium picolinate and carnitine, an amino acid that helps prevent fats from building up within the body. The total daily intake of chromium during this phase was 600 micrograms.

At the end of this phase, an average reduction in weight of 15.1 pounds was noted, accompanied by an average loss of

What Is a Calorie?

If you've ever gone on a diet, you may have spent a bit of time counting your calorie intake. But what are you counting, exactly?

Calories are used to measure the amount of energy given off when a certain amount of food is burned. But individual calories represent tiny amounts of energy, so it is easier to talk about food energy in terms of kilocalories—1,000 calories taken together. Thus, when we say that a cookie contains 60 "calories," we really mean that it contains 60 "Calories," or 60,000 little calories. For simplicity's sake, we will use the term "calories." Just keep in mind that what we call "calories" are really kilocalories, or "Calories."

11.8 pounds body fat. This means that most of the weight loss consisted of fat loss, instead of loss of either water or muscle. Total cholesterol levels decreased by 11 percent and LDL cholesterol levels decreased by 9 percent during this phase. Metabolic rate increased by 21 percent when the participants ate the cookies and took the supplements.

The results of Kaats's second study help explain why subjects were able to lose fat without exercise and without nearly starving themselves, since quick weight loss usually requires a diet of less than 800 calories a day. The participants' body cells were actually using more energy, so they were burning more fat. Most often, fat-reduction plans cause a marked decrease in cell activity for two reasons: the muscles aren't getting enough energy from the diet, and stored fuel—body fat—doesn't get transferred to the muscle cells. When there's enough chromium in the body to make the insulin work efficiently, fat, a high-quality fuel, gets used for energy. This means that the cells are burning more energy than they would otherwise. (See Chapter 6 for more information on chromium picolinate and fat loss.)

The second Kaats study also demonstrated the importance of fiber in the diet. Health professionals have been encouraging us for years to increase our dietary fiber. Fiber gives a diet bulk. This helps satisfy hunger, especially when the amount of food eaten is reduced. Fiber may also slow absorption of sugars and fats from the diet, so that the incoming fuel might be used for energy rather than being stored away as fat, which is what happens when all the fuel comes rushing in at the same time.

Another important part of the Kaats' study was the use of carnitine. Carnitine is an amino acid you don't hear much about. It is absolutely essential for moving fats into the furnaces of the cells, called the mitochondria, where the fats are burned and energy is released. The mitochondria are the only parts of the cell where fat can be broken down into energy. Without carnitine to act as a carrier, fat cannot get into the mitochondria.

Like picolinate, carnitine can be made in the body, so scientists disagree a lot about whether or not supplemental carnitine needs to be added to diets. Based on my study of the scientific literature, I am convinced humans cannot get enough raw material from the diet to make a sufficient amount of carnitine. Muscle cells require a generous amount of carnitine to get all the needed fat into the mitochondria.

In a study conducted by the Department of Medicine at Karl Franzens University in Austria, a team of doctors led by Dr. B. Bahadori found that chromium picolinate prevents the muscle deterioration that often occurs in people on weight-loss diets.[9] All the people in this study were first fed a very-low-calorie diet—800 calories a day—for eight weeks. As expected, by the end of that period everybody had lost weight, about fifteen pounds on average.

The participants were then divided into four different groups: a normal diet, a high-fiber diet, a high-fiber diet supplemented with chromium in the form of yeast, or a high-fiber diet supplemented with chromium picolinate. Each supplement contained 200 micrograms of chromium.

After four months, lean body mass decreased in all the people except those given the chromium picolinate. In those people, muscle mass actually increased by nearly four pounds. This study shows that much of the weight lost on diets is often not fat—it is actually muscle and some water. The doctors who conducted the study suggested that people on weight-reduction diets should also take chromium picolinate to prevent muscle loss.

REDUCING BLOOD SUGAR IN DIABETICS

One of the most exciting reports about chromium came to me in the form of an unsolicited package from Dr. A. Ravina of Haifa, Israel. It included a letter of introduction along with the English translation of a journal article. The letter described how chromium picolinate was being advertised in Israel as being "good for diabetics." This prompted patients to ask their doctors about the product, which in turn prompted the doctors to conduct clinical trials. The article described the results of such a study.[10]

The article described observations of 162 diabetic patients. Of these, 48 could not produce insulin at all—they were insulin dependent. The other 114 could produce insulin, but it didn't work efficiently—they were non-insulin dependent. (See Chapter 7 for a more detailed discussion of diabetes.)

At the beginning of the study, the doctors measured how well the patients responded to insulin by injecting insulin into each patient and then measuring how fast it cleared glucose, or blood sugar, from the bloodstream. Each patient was then given ten chromium picolinate capsules containing 200 micrograms of chromium each, and told to take one capsule at breakfast for the following ten days. Most of the diabetics in the second group, those who were non-insulin dependent, were being treated with some type of drug. So the doctors had these participants cut down on their drug dosages to keep their blood-sugar levels from dropping too low if the chromium supplement worked. After ten days, the patients returned for a second insulin response test.

Of the 28 insulin-dependent women, 71 percent showed a positive response. Of the 20 men in this group, 70 percent responded positively. Of the 63 non-insulin-dependent women, 73 percent showed a positive response. Of the 51 men in this group, 74 percent responded positively. A positive response, of course, was an improvement over the initial insulin response test. Keep in mind that these patients had been using chromium picolinate supplements for only ten days! Also included in this study were some patients who were able to control their blood glucose more efficiently while using chromium picolinate, but who reverted to high blood-glucose levels a few days after the chromium was discontinued.

Many women have higher-than-normal blood-glucose levels during pregnancy. This condition is called gestational diabetes, and is a real concern for both the doctor and the prospective mother because the condition can, and sometimes does, lead to permanent diabetes.

A team of doctors, led by Dr. Lois Jovanovic-Peterson at Sansum Medical Foundation in Santa Barbara, California, studied the effect of chromium picolinate on 30 pregnant women, beginning in the 20th to 24th week.[11] Ten were given 4 micrograms of chromium for every kilogram, or little over two pounds, of body weight. Ten were given 8 micrograms per kilogram, and ten were given a placebo. The study lasted eight weeks.

The women given chromium picolinate had significantly lower levels of insulin, blood glucose, and glycated hemoglobin, a chemical found in the blood of diabetics (see Chapter 1).

Compared with the women who received the placebo, the women taking the supplement needed approximately half as much insulin to clear glucose from their bloodstreams, and their average blood-sugar level was 20 percent less.

At the end of the study, each woman was given a drink that contained glucose. An hour later, her blood was tested. The women who were given the smallest amount of chromium picolinate had much lower blood-glucose levels than the women who were given the placebo. But the women who were given the greatest amount of chromium picolinate had blood-glucose levels that were about the same as those of nondiabetic women. This indicates that chromium picolinate can completely reverse the development of gestational diabetes.

These studies show that chromium picolinate improves the efficiency of insulin. This means that blood glucose can be regulated effectively without the secretion of excessive insulin often encountered in pregnant women. Insulin is a valuable hormone, but the body doesn't function well when too much insulin has to be put into the blood to control blood sugar (see Chapter 5).

Overall, the studies in this chapter supported the results we saw in the studies described in Chapter 1.

3. PIGS DON'T LIE—
ANIMAL STUDIES

U p to this point, we have discussed chromium picolinate studies that involved human beings, using terms
such as "placebo" and "double-blind." (For defini-
tions of these terms, see "The Science of Studies" on page 7.)

Human studies require these kinds of precautions because
the human mind can often influence the human body. People
in a study will often respond positively just because somebody is paying attention to them—what scientists call the
placebo effect. It is important that a study's participants don't
know whether they are taking the test substance or the inert
placebo pill. This way, they won't do something to cause the
results to change. The need for these precautions makes quite
a statement—that humans are the least desirable experimental subjects.

That's why I can't stress enough the value of the animal
studies in this chapter. These studies—with pigs, rats, and
other animals—have given us insight into how important
chromium picolinate is and what it does in the body.

After discussing the studies themselves, we'll take a quick

look at a way to compare the amount of chromium needed by various species, including human beings.

STUDIES WITH PIGS: MORE EVIDENCE
OF CHROMIUM PICOLINATE'S BENEFITS

Animal scientists are constantly searching for inexpensive and convenient methods to increase the quantity and quality of farm products. This is particularly true in the swine industry, where the goal is to produce leaner pork. Drugs and hormones have been used successfully for that purpose, but they are not in common use because of a lack of acceptance by consumers and a lack of approval by the Food and Drug Administration.[1] Chromium picolinate offers the industry a promising option.

Studies with pigs have concentrated on both the need for leaner, meatier pork and on the effect of chromium picolinate on reproduction.

More Meat, Less Fat

In the late 1980s, Tim Page, a graduate student at Louisiana State University (LSU), became interested in dietary nutrients as a way to improve the quality and quantity of pork. He was aware of how essential chromium is to good health, and in fact had done some unsuccessful studies with chromium salts. After learning of our studies with chromium picolinate, Page decided to devote his doctoral research—the research that earned him his degree—to studies with chromium picolinate.[2-7] His work has been very valuable to both animal scientists and human health professionals.

In several of Dr. Page's studies, chromium picolinate was added to the regular feed of pigs from the time they were weaned until the time they were ready for slaughter. These pigs were compared with pigs that were not fed chromium picolinate. Different animals received different amounts of chromium picolinate so the scientists could get an idea of the amount needed to get the best results.

After the pigs were slaughtered, the amount of fat along the back, called the backfat thickness, was measured along with the amount of muscle in the body overall and in the loin eye area, where pork chops come from. Even the pigs given the smallest amount of chromium picolinate showed improvement over the pigs that didn't get any chromium. But pigs fed the two largest amounts of chromium picolinate showed the best results—backfat thickness decreased by 30 percent, overall muscle increased by 14 percent, and the loin eye area increased by 23 percent—compared with pigs that were not fed supplements.

The possibility that an essential nutrient, rather than some undesirable drug, could have such noticeable effects intrigued the scientists at LSU. As they continued their studies, the outcome was always the same—the use of chromium picolinate supplements consistently resulted in pigs with less fat and more lean meat.

In one study, the LSU scientists used pigs from a completely different family so they could be sure that the results they were seeing were not due to some odd genetic trait. These pigs had an average of 20 percent less fat and 7 percent more muscle. Thus, the scientists learned beyond a doubt that their findings were due to the chromium picolinate.

When the LSU scientists compared the effects of chromium picolinate with those of chromium salts, they obtained some astonishing results. In one experiment, pigs eating chromium in the form of chromium salts lost only 5.5 percent of their fat and didn't gain any muscle. But pigs fed an equal amount of chromium in the form of chromium picolinate lost 22 percent of their body fat and gained 6.5 percent more muscle. Suspecting that the salt form was not as effective as the picolinate, the researchers added much more chromium to the diet of the pigs fed chromium salts. Despite the fact there was 300 times more chromium in salt form than in picolinate form, no response could be elicited using chromium salts—proving that chromium picolinate is the most effective way to get chromium into the body.

Just to be sure the responses seen in his experiments were not caused by the picolinate itself, Dr. Page and his associates added picolinate alone to the diet of one group of pigs. There was no change in either fat or muscle content, proving that picolinate without chromium does not lead to decreased fat or increased muscle.

Another thing the LSU researchers noticed was that in three of six studies, blood-cholesterol levels decreased by between 5 and 20 percent in pigs fed chromium picolinate.[8,9] It's difficult to translate these results into human terms, since a pig has about half the total cholesterol of a person. However, USDA researchers noted that chromium picolinate fed to pigs resulted in an increase in the amount of good, system-cleaning HDL cholesterol over the amount of bad, system-clogging LDL cholesterol.[10]

The LSU studies have been verified at Oregon State University and the University of Kentucky. One study showed 10 percent less fat, while the other showed 12 percent less fat—and 6 percent more muscle—in pigs fed chromium picolinate daily.[11,12]

These results have also been verified by researchers working in other countries. For example, in a study conducted at the National Institute of Agriculture in Taiwan, three groups of pigs ate three different amounts of chromium picolinate.[13] Compared with pigs given no supplement, the pigs eating the smallest amount of chromium picolinate had 16 percent less fat. Twice as much chromium picolinate resulted in 26 percent less fat, and three times as much resulted in 32 percent less fat.

Chromium Picolinate and Pregnancy

Animal studies at another university verified another aspect of the human studies we discussed in Chapter 2. These animal studies, done by Dr. Merlin Lindemann and his associates at Virginia Polytechnic Institute (VPI), showed that chromium picolinate not only reduced the body fat in pigs, but also improved the action of insulin.[14,15] The VPI researchers extended their studies, and got results that are not only important to the swine industry, but that also have implications for human health.[16]

Lindemann's team fed female pigs chromium picolinate from early youth through the time they had delivered two litters. Sows fed chromium picolinate for more than nine months used less insulin to clear glucose from their bloodstreams than those fed chromium picolinate for thirty-five days, which in turn used half as much insulin as sows that were not fed any chromium picolinate. Without the chromium picolinate supplement, insulin became less and less effective as the pigs matured. (See Chapter 5 for more information on the action of insulin in the body.)

In addition, chromium picolinate had an effect on the health of the piglets. The number of piglets born alive and the number of piglets alive after twenty-one days were both 26 percent higher for sows given chromium picolinate than for sows that were not fed supplements, and the litter weight, or total weight of the litter, of the piglets born to supplement-fed sows was 26 percent greater at birth and 22 percent greater after twenty-one days. Of the 11 sows that did not get chromium picolinate, only 6 had second litters. But 10 of 11 chromium-fed sows had second litters.

In another experiment, the VPI investigators fed female pigs chromium picolinate daily until they were bred, at which point the supplements were stopped. The idea was to determine if the chromium picolinate would have any carry-over effect on the piglets even after the sow had stopped taking it. It did. The number of piglets born alive to these sows and the number of piglets alive after twenty-one days increased by 13 percent compared with piglets from sows that were not given the supplement. The litter weight of the piglets from the supplement-fed sows was 18 percent greater at birth and 21 percent greater at twenty-one days.

STUDIES WITH RATS: FAT LOSS AND LONGER LIFE

Rats, long used in medical studies, have been used in chromium picolinate research. Rat studies have verified the fat-loss effect noted in studies with pigs and with humans. But some of the most interesting rat studies have revealed a link between chromium picolinate and longevity.

Rats Lose Fat and Gain Muscle

Pigs are not the only animals that lost fat and gained muscle when given chromium picolinate. Deborah Hasten of LSU—the researcher we met in Chapter 2—and her associates produced leaner and healthier rats with chromium picolinate supplements.[17,18] In one study, rats were fed chromium picolinate daily and made to run three times a week. Other rats were also made to exercise, but were not fed chromium picolinate. After twelve weeks, the rats fed the supplement gained about 15 percent more leg and heart muscle than rats given no supplement.

In another study by Hasten,[19] different rats were fed different amounts of chromium picolinate for ten weeks. Rats given no chromium picolinate had 19.8 percent body fat at the end of the study. The rats fed the smallest amount of supplement had 17.5 percent body fat, a reduction of 11.5 percent. Rats fed the intermediate amount of supplement had 15.4 percent body fat, for a reduction of 22 percent, and the rats fed the highest amount of supplement had 13.4 percent body fat, a reduction of 32 percent compared with rats fed no supplement.

Rats and Longevity

So far, we have seen how many good things chromium picolinate can do in the body, everything from reducing levels of sugar and fat in the bloodstream to reducing body fat. These effects all stem from chromium's ability to help insulin do its job properly (see Chapter 5). But, in addition to all these benefits, there is evidence that chromium picolinate can help provide the ultimate benefit: a longer, healthier life.

I mentioned glycated hemoglobin in Chapter 1, in the discussion on chromium picolinate and diabetes. Glycation happens when blood sugar, or glucose, attaches to various proteins and other substances in the bloodstream. One of the most common glucose-protein combinations is the attachment of glucose to hemoglobin, the chemical that allows the blood to carry oxygen. Glycation is most common in diabetics but it occurs whenever glucose levels become higher than Nature intended.

In the mid-1980s, a scientist interested in the aging process thought that glycation might be the primary cause of aging.[20] To test this idea, a group of researchers at Texas A and M University fed rats a diet that contained 60 percent of the calories rats normally eat.[21] This calorie restriction was intended to reduce the amount of glucose in the rats' blood. calorie restriction had first been tested in the 1930s, when it was discovered that rats fed calorie-restricted diets lived much longer,[22] a result confirmed by other studies.[23] What worked in the 1930s worked fifty years later; the Texas group found that their rats not only lived longer than rats fed a normal diet, but that the rats also had lower blood-glucose levels and much less glycated hemoglobin.

The results published by the Texas scientists interested me because we had just completed the study with diabetics at Mercy Hospital described in Chapter 1. After six weeks of chromium picolinate supplements, diabetics had lower levels of blood glucose and glycated hemoglobin.[24] I also remembered a description of studies showing how addition of chromium to diets extended the life of rats and mice.[25]

As a result, I and my colleagues at Bemidji State University (BSU) decided to see if chromium added to the diet of rats would increase their life spans.[26] We wanted to know if the type of chromium used would make a difference in the results, so we divided a group of recently weaned rats into three smaller groups, and gave each group a different form— chromium picolinate, chromium salts, or a chromium-niacin combination—of the same amount of chromium. The rats were allowed to eat as much as they wanted.

After 1,000 days, the blood-sugar levels of the rats fed chromium picolinate were about 20 percent lower than those of the rats fed either chromium salts or the chromium-niacin combination. Glycated-hemoglobin levels in rats fed chromium picolinate were 40 percent of those found in the other rats.

We also measured the rats' insulin levels. The lower the concentration of insulin in the bloodstream, the longer the rats survived. Insulin concentrations in rats fed chromium picolinate were about 50 percent less than those in rats from the other two groups. This means that the rats eating the chromium picolinate were using insulin more efficiently than the other rats.

The rats fed chromium picolinate lived longer than the rats fed either chromium salts or the chromium-niacin combination. The rats fed chromium picolinate lived an average of 1,316 days, while the rats from the other two groups lived an average of 1,033 days—a difference of more than nine months, or a considerable part of an average rat's life span. At death, the rats fed chromium picolinate not only weighed less, but had an average body-fat level that was 26 percent less than that of the other rats.

This study took four years to complete, but was well worth the effort because it gave us a lot of valuable information about chromium in the diet. It did show that rats eating the chromium picolinate lived longer than the rats in the other two groups. The rats getting the other, less available forms of chromium didn't look sick, but they weren't getting enough chromium. This caused higher blood-glucose levels and earlier death.

The experiment also showed that the increased survival rate in rats fed the chromium picolinate was probably related to how well their insulin was working, since they had an average blood-insulin level that was one-half of the level found in the other rats. Throughout their lives, glucose levels in rats fed chromium picolinate were 15 to 20 percent lower than in the rats in the other two groups, and there were lower levels of glycated hemoglobin. These lower levels of blood glucose and glycated hemoglobin meant that insulin was working better, and it is properly functioning insulin that is an important factor in slowing the aging process. For a more detailed discussion of chromium picolinate and its effects on aging, see Chapter 9.

STUDIES WITH OTHER ANIMALS: FURTHER PROOF

While most chromium picolinate animal studies have been done with pigs and rats, research has been done with other animals. Most of these studies are valuable in that they confirm the results seen in the experiments we've examined so far. But several studies show that chromium picolinate can provide an unexpected benefit for human beings—low-cholesterol eggs.

Verifying the Benefits of Chromium Picolinate

Experiments with chromium picolinate supplements in animals consistently result in decreased body fat and increased muscle. Investigators at LSU found less body fat in growing chicks[27] and lambs[28] when chromium picolinate was added to the feed. Animal studies have also verified chromium picolinate's ability to reduce blood cholesterol. In two experiments with calves, researchers from LSU found from 20 to 26 percent less total cholesterol in calves fed chromium picolinate than in calves that ate a regular diet.[29]

The link between muscle development and chromium picolinate was verified in an experiment done by a veterinarian in Conifer, Colorado. Dr. H. C. Gurney, Jr., proved that chromium picolinate helps prevent or reverse muscle deterioration in dogs.[30]

Lame dogs were used in this three-year study because lameness results in muscle atrophy, or breakdown. The group of 167 dogs were divided into smaller groups based on the severity of each dog's muscle atrophy. The amount of chromium picolinate each dog received daily was based on its weight—dogs that weighed less than twenty pounds were given 100 micrograms, dogs that weighed from twenty to fifty pounds were given 200 micrograms, and dogs that weighed more than fifty pounds were given 400 micrograms. Each dog's owner was asked to rate the animal's condition as better, the same, or worse, based on how well the dog could stand up without help and how well it could walk, and on the dog's endurance.

As expected, the dogs with the most severe muscle atrophy showed the least improvement. Those that did show improvement—8 of 27—needed forty-five to ninety days of supplements before the improvement was noticeable. Among the dogs with less severe atrophy, 71 of 104 improved within only fifteen to thirty days. In dogs that were obese and had lost the ability to stand up without help, 6 of 9 showed improvement after thirty to forty-five days of supplements.

This study shows that chromium picolinate helps maintain and develop muscle. As Dr. Gurney points out, many dogs with muscle weakness and atrophy are treated with steroids that produce adverse effects. Without doubt, if dogs respond to supplements of chromium picolinate, humans with muscular weakness should also respond. Chromium picolinate supplements would be wise for people who are or have been bedridden.

Low-Cholesterol Eggs?

The food eaten by laying hens contains very little chromium. Because chromium supplements can decrease blood cholesterol, Dr. C. Y. Hu at Oregon State University and his colleagues from the National Institute of Agriculture in Taiwan tested the possibility of lowering egg cholesterol by feeding hens chromium picolinate.[31] The amount of cholesterol in the yolk from eggs produced by supplement-fed hens was reduced by as much as 33 percent compared with eggs from hens given no chromium picolinate, depending on how much chromium picolinate was used. These studies show that it is possible to produce eggs with less cholesterol.

An equally exciting result of Dr. Hu's studies was the change in blood-cholesterol levels shown by the hens given chromium supplements. Levels of the good HDL cholesterol increased by between 28 and 70 percent! At the same time, levels of the bad LDL cholesterol decreased by between 25 and 69 percent.

Dr. Page—the researcher who did many of the pig studies we discussed earlier—and his associates at LSU also tested the possibility of lowering egg cholesterol by feeding hens chromium picolinate.[32,33] Unlike Dr. Hu and his colleagues, the LSU scientists didn't produce low-cholesterol eggs. However, they did find lower blood-cholesterol levels in the hens given chromium picolinate. An unexpected result of the LSU experiments was an increase in egg production from hens given chromium picolinate.

COMPARING CHROMIUM AMOUNTS ACROSS SPECIES

It is not always easy to translate the results of studies done with pigs, rats, and hens into amounts of chromium picolinate required by human beings. That's because there are marked differences in size and metabolic rate between species. For example, a rat, being smaller and having a higher metabolic rate than a person, needs a different amount of chromium than a person.

One way to overcome this problem is to look at the number of calories consumed. This way, you don't have to know exactly how much food an animal eats each day. All you have to consider is the amount of chromium and the number of calories in a certain amount of food. This chromium-to-calorie ratio, which I call the Chro Cal, allows us to compare chromium amounts across species.[34]

With the Chro Cal, we can compare the results of experiments on pigs, rats, chickens, or humans to determine if different species have different chromium requirements. Take, as an example, the pig studies described earlier. The average number of calories was 2,950 per kilogram—or a little over two pounds—of feed. Very good results were obtained when the feed contained from 200 to 400 micrograms of chromium per kilogram, or between 68 and 136 Chro Cals.

The similarities among the studies with humans, pigs, rats, and other species are absolutely astonishing. In pigs, eating between 130 and 250 Chro Cals led to a fat reduction of between 25 to 32 percent. In rats, consumption of 150 Chro

Cals reduced fat by 24 percent. The use of between 150 and 250 Chro Cals produced the most consistent reductions in body fat.

The studies with rats and pigs were started when the animals were young and continued until they reached maturity. All the human studies were done with adults and lasted from six to twelve weeks, making direct numerical comparisons with the rat and pig studies impossible. However, the relative outcomes were the same. During the studies at BSU, the athletes ate about 70 Chro Cals and lost about 22 percent of their body fat, about the same as what was noted with rats and pigs.

The same similarities exist when you look at blood cholesterol. Chro Cal levels of about 70 to 85 resulted in less LDL cholesterol and more HDL cholesterol in pigs, hens, and people.

We can say that the best results in terms of body-fat reduction, increased muscle development, lower cholesterol, and increased insulin efficiency can be obtained with approximately 150 to 250 Chro Cals. The use of chromium picolinate has resulted in these benefits in every species, including human beings, it has been tested in.

Assuming the average person eats about 2,500 calories a day, the daily intake of chromium should be between 400 and 600 micrograms of chromium a day. Most chromium picolinate products contain 200 micrograms of chromium, so two to three capsules or tablets a day should provide the chromium your body needs to start functioning at maximum capacity. If you have a specific concern, such as obesity or diabetes, see Chapters 6 through 9 for more detailed recommendations.

4. WHY YOUR BODY NEEDS CHROMIUM PICOLINATE

Simply put, you need chromium picolinate because chromium is an essential mineral that is in short supply in food, and picolinate is the substance that can get chromium into the bloodstream.

In the mid-1980s, the United States Department of Agriculture (USDA) did a study that showed how inadequate our diets are in providing us with enough chromium.[1] None of the volunteers, who were asked to record what they ate for a week, ate more than 33 micrograms of chromium a day. This was less than the official recommended chromium intake of between 50 and 130 micrograms a day.

I hope that the studies discussed in the first three chapters of this book have convinced you both of the body's need for chromium and of picolinate's ability to make chromium available within the body. In this chapter, we'll look at how scientists came to understand how important chromium is and why scientists don't agree about the importance of this mineral. We'll then look at what clues led me and others to discover why picolinate is the key to chromium nutrition.

HOW SCIENTISTS DISCOVERED
THE NEED FOR CHROMIUM

The seeds for studies into chromium's importance began to germinate during the 1920s, when researchers began to understand the nature of diabetes. But they were slow to develop, and for many years scientists did not recognize how important chromium is to human health.

Clues From Insulin Studies

In the early 1920s, Dr. Frederick Banting and his assistant, Charles Best, working at the University of Toronto, proved that a substance produced by the pancreas—which they called insulin—could be used to arrest the symptoms of diabetes, a disease that has afflicted humanity throughout history. (For more information on diabetes, see Chapter 7.) Later in the decade, more research showed that insulin was much more effective when it was mixed with a yeast extract.

In the late 1950s, Dr. Klaus Schwarz and Dr. Walter Mertz discovered chromium's importance to health almost by accident.[2,3] In experiments with the mineral selenium, they had been feeding their rats diets supplemented with Torula yeast, a source of needed protein.

While doing standard blood tests on these rats, Schwarz and Mertz discovered that blood from the rats eating the diets with Torula yeast contained more glucose, or blood sugar, than was considered normal. This meant that the rats fed Torula yeast could not remove glucose from their bloodstreams as quickly as they should have. The rats had a condition known as glucose intolerance, which most often occurs when insulin isn't working efficiently.

Schwarz and Mertz had read reports describing how extracts from brewer's yeast, a beer-making byproduct, improved insulin action. So they added brewer's yeast extract to the rats' diets, with the result that the rats were able to clear glucose from their bloodstreams normally. These studies showed that brewer's yeast had something apparently lacking in the Torula yeast. The pair named the unknown substance the glucose tolerance factor (GTF). They later discovered that pork kidney extract could also be used to prevent glucose intolerance in rats.

Exactly what was the GTF? After a lot of hard work and careful analyses, Schwarz and Mertz discovered that the brewer's yeast and the pork kidney extract had something in common—they both contained chromium. Because they did not find chromium in Torula yeast, they suspected that chromium was one of the ingredients in the GTF.

The pair also strongly suspected that the chromium in these supplements was connected to some other substance, but were never able to confirm their suspicions. This failure to discover the nature of the other ingredient contributed to later skepticism about chromium's importance as a nutrient.

Clues From Mineral Studies

The idea that chromium could be of such importance intrigued many researchers, who began trying to confirm the findings of Schwarz and Mertz. One of these scientists was Dr. Henry Schroeder, a researcher we first met in Chapter 1. He conducted many experiments[4-6] with rats and mice that showed chromium is needed to help control the amounts of both glucose and cholesterol in the blood. He also described studies in which rats and mice lived longer when chromium was put into their drinking water.

Dr. Schroeder was the first scientist to link low chromium intake with heart disease. He carefully analyzed body organs taken from people who died of heart disease. When he compared these organs to organs from people who had died accidentally, he discovered that people with heart disease had much less chromium in their bodily tissues. Dr. Schroeder and his colleagues also found less chromium in Americans than in people from third-world countries, where the incidence of diabetes and heart disease is very low.

A series of interesting studies conducted by Dr. Schroeder and his group involved measurement of chromium in the organs of people as they aged. This study showed that chromium is fairly plentiful in the human body shortly after birth, but that the amount decreases as a person grows older. Equally interesting was the fact that women who had not had children had more chromium in their bodies than women who had children, while women who gave birth to only one child had more chromium than women who had borne two children, and so on. These studies tell us babies get chromium by taking it from their mothers while they are in the womb, but after the umbilical cord is cut they don't take in enough chromium to counteract losses through the urine. This happens because most diets don't have enough chromium, and what chromium exists can't be absorbed and used effectively. Dr. Schroeder's group also showed that food lost chromium during cooking.

Accumulated Proof and a Very Sick Woman

In 1969, Dr. Mertz published an article about all the experiments done with chromium up to that time.[7] In the first chromium study ever done with humans, no improvement in blood-sugar control could be seen in 7 adult diabetics who were given a daily chromium supplement for one week. However, in another trial, six diabetics supplemented their daily diet with chromium. By the end of the four-month trial, 3 of the 6 showed marked improvement in blood-glucose control. I have no doubt that all six patients would have eventually improved had the study lasted longer.

Another study described by Dr. Mertz showed that undernourished children respond much more rapidly to chromium supplements than adult diabetics. The children were given a single juice drink containing chromium. After only eighteen hours, many of them showed great improvement in their ability to clear glucose from their bloodstreams. While this study's results have been described as questionable, it is important to realize that many of the participants showed improvement after only one chromium supplement. Undoubtedly, circumstances prevented the scientists who did the study from continuing it. But the results they did obtain were impressive, especially since children who were not given chromium showed no improvement.

Some of the chromium studies described in Dr. Mertz's article include some intriguing observations. One study showed that chromium levels in the bloodstream increase after a person eats. This finding tells us that chromium is moved to the bloodstream from storehouses in the tissues while food is being digested and absorbed. Since insulin is secreted when food is digested, one can assume that chromium and insulin work together to remove glucose from the blood.

In the late 1970s, twenty years after Schwarz and Mertz first linked chromium to proper insulin action, three studies provided unquestionable evidence that the body needs chromium to control blood sugar. In these studies, the value of chromium was discovered quite by accident.

In the first study, Dr. K. Jeejeebhoy and his colleagues at Toronto General Hospital were treating a middle-aged woman who couldn't digest or absorb the food she ate.[8] She had to be fed through a needle inserted into a vein. This intravenous feeding is called total parenteral nutrition (TPN).

After more than three years of TPN, the woman developed diabetes symptoms—a very high blood-sugar level, weight loss, loss of muscle movement control, and loss of feeling in her limbs. Fortunately for that woman and millions of people after her, one of the doctors treating her was familiar with the chromium studies we've just discussed. Chromium was added to the TPN fluid, and two weeks later she had normal blood-sugar levels, her muscle control was restored, and she had gained back the lost weight.

Two years after this landmark discovery, another patient on TPN developed high blood-sugar levels after five months of therapy. Chromium was added to the TPN fluid, and three days later, the patient's blood-sugar levels returned to normal.[9]

High blood-sugar levels, loss of glucose through the urine, and weight loss appeared in a third patient who had been on TPN therapy for seven months. Again, shortly after chromium was added to the intravenous fluid, the amounts of blood and urine glucose dropped back to normal, and the patient began gaining weight.[10]

These successful treatments, coupled with the dozens of experiments with both humans and animals, forged a strong link between chromium and insulin action.

WHY SCIENTISTS DON'T AGREE
ABOUT THE IMPORTANCE OF CHROMIUM

In 1980—mainly because of the results published by Jeejeebhoy's group—the National Research Council (NRC), a branch of the National Academy of Sciences, adopted what is known as a "safe and adequate daily intake" for chromium of between 50 and 200 micrograms. Despite this endorsement— one made without much enthusiasm by the NRC—chromium didn't gain a lot of acceptance as one of the essential nutrients. During the 1980s, fewer and fewer scientific studies on chromium were being done. In 1988, an editorial in one prestigious journal raised doubts about the need for chromium in the diet. It questioned whether or not chromium had met the standards for essentiality.[11]

What are these standards? Dr. Mertz[12] summarizes them:

- The nutrient must be in living matter.
- The nutrient must interact with living systems.
- A dietary deficiency of the nutrient must consistently cause a bodily function to fail, at least partially.
- This failure in function must be preventable or curable by adding useful amounts of the nutrient to the diet.

Let's consider why many health professionals claim that chromium has not fully and convincingly met these standards.

Chromium Levels Are Low Everywhere

The first standard, that the nutrient be found in living matter, is met because chromium is indeed present in living cells. But the amount found in soil, plant and animal tissues, and blood is very low. This makes detection of chromium extremely difficult. Blood chromium analyses require meticulous handling of samples and sophisticated, expensive equipment.

Contamination is another problem in chromium analyses because chromium is used to make needles and other medical supplies. Shortly after Schwarz and Mertz discovered the essentiality of chromium in animals, many investigators started analyzing food and bodily tissues for the mineral, but didn't take necessary precautions against contamination. As a result, most early published values have been shown to be wrong, giving skeptics reason to question the meaning and validity of early experiments with chromium.[13]

Where Are Chromium's Chemical Partners?

The second standard, that the nutrient must interact with living systems, may turn out to be the most difficult to fulfill.

Copper, zinc, iron, and even selenium have been found to be associated with important enzymes or other proteins in blood or tissue. Chromium has been found in blood and cells, but hasn't been associated with any specific part of the cell or any cellular chemical. Until that happens, critics will claim that the observations on chromium and insulin don't prove that chromium is essential.

Nutrients that can't be made by living cells but must be obtained from the soil have always been slowest in gaining acceptance by health professionals. Iodine stands out as a prime example.

For centuries, millions of people suffered from a disease called endemic goiter, caused by iodine deficiency. As early as 1820, a French doctor was using iodine to treat goiter. Twenty years later, a French botanist showed that goiter occurred in areas where the iodine content of the soil was low. He even went so far as to suggest iodine be added to the water in areas where goiter was prevalent. But he was ignored, and the idea was discredited. Iodine was forgotten.

Iodine wasn't rediscovered until 1919, when scientists studying the thyroid gland found that iodine is required for that gland's proper functioning. When iodine supplementation was started, goiter and its related health problems were virtually eliminated.

Iodine and chromium have much in common. Both must be taken into the body from the diet, both are often lacking in the food we harvest and eat, and both are needed by vital hormones in the body. When iodine is deficient in the food supply, the gland requiring it enlarges. When chromium is in short supply, the pancreas doesn't enlarge but instead produces too much insulin.

Iodine and chromium are also similar in the effort required before either nutrient was determined to be essential. Decades before iodine was detected in the thyroid, doctors were using iodine supplements to cure and prevent goiter. However, their results were dismissed as questionable. Today, successful trials with chromium supplements are questioned because nobody has found a part of the cell that depends on chromium. I predict that one day somebody will find a way to analyze insulin just before it combines with a cell, and will find that the most efficient form of insulin will have chromium attached to it.

What Is Chromium's Purpose Within the Body?

Some skeptics claim that chromium doesn't meet the third standard for essentiality—that a diet deficient in the nutrient must consistently cause a bodily function to fail, at least partially—because chromium depletion doesn't always produce the same, noticeable effect. These people argue that chromium's job within the body must be identified before it can be accepted as essential.

We know that for insulin to function effectively, it needs chromium. So exactly how does insulin function? I provide a more detailed explanation in Chapter 5, but briefly, insulin works by first hooking onto the outside of a cell. This causes a chemical message to be sent to the inside of the cell, which summons specialized chemicals designed to remove glucose from the bloodstream. This call to the inside of the cell is a crucial step in the process of clearing glucose from the blood. If insulin can't relay the message, blood-sugar levels remain high, causing more insulin to be secreted.

Humans diagnosed as chromium deficient have many impaired bodily functions, from reduced glucose tolerance to high levels of blood cholesterol. These malfunctions can be linked to one specific cause—poor insulin function. However, because chromium has never been found in insulin or any of the cell parts that bind with insulin, critics feel there isn't enough evidence to establish chromium's essentiality in humans.[14]

We don't know exactly how chromium interacts with insulin. But we do know from laboratory studies that cells respond much more effectively to insulin when chromium is present. In these experiments, fat cells are taken from an animal and put into test tubes containing the fluid needed to keep the cells alive. When chromium is added to the fluid along with insulin, the action of insulin is much more efficient because glucose enters the fat cells at a greater rate.[15] These studies show that chromium does not have to be built into either the insulin or the cell in order to have an effect. Chromium's presence in the blood or cell fluid somehow makes insulin's job a lot easier.

Chromium Supplements Don't Always Work

Chromium's inconsistency in meeting the fourth standard of essentiality, that failure in a bodily function must be preventable or curable by useful amounts of the nutrient, has been a key reason why chromium hasn't been readily accepted as an important nutrient for humans. Complying with that standard requires causing a deficiency in a test subject and then curing and preventing the symptoms by putting the nutrient in question back in the diet.

Without sophisticated equipment, and given the difficulties of chromium analysis, many investigators simply assumed that diabetics and people with high blood-cholesterol levels didn't have enough chromium, and therefore didn't do the necessary tests. In many studies, patients with high levels of blood sugar or cholesterol were given chromium supplements in an attempt to regulate these problems.[16]

Unfortunately, most of the patients were given chromium chloride, or chromium salts. In chromium chloride, the positively charged chromium is attracted to negatively charged chloride. When the salt is put into fluid, such as the fluid within the intestines, the chromium and chloride are pulled apart, leaving the charged components floating around by themselves. We now realize that cells repel charged chemicals, so it's not surprising to learn that many of the attempts to control blood glucose or cholesterol with chromium salts met with failure.

Some investigators, recalling the original experiments of Schwarz and Mertz, tried to supply a more natural source of chromium by giving the volunteers yeast that had been grown with high amounts of chromium.[17–19] This approach was more successful. But not all health professionals were convinced it was the chromium that was causing the good results because yeast contains not only chromium, but many other vital nutrients. The use of yeast also didn't gain much popularity with the general public, because too much yeast—from 50 to 100 grams a day—is needed to get favorable results. Also, many people are—or at least think they are—allergic to yeast.

The unpredictable results with chromium supplements made many scientists skeptical about the true value of chromium. The inability to identify the exact nature of the GTF reinforced that skepticism. In 1974, Dr. Mertz's research group at the USDA presented a series of papers in which they argued that the GTF consisted of chromium chemically bound to the vitamin niacin, and to either an amino acid or a chemical made of amino acids.[20] When this substance, the alleged GTF, was made synthetically, it did indeed increase insulin efficiency. However, the substance was not stable enough in bodily fluids to be useful. On the other hand, the yeast and kidney extracts worked just fine. Why?

THE PICOLINATE CONNECTION

So far, we've seen that although the human body needs chromium, coming up with a chromium supplement was a difficult task. Now we need to find out more about chromium's ideal partner, picolinate.

Chromium Needs To Be Smuggled Inside

Chromium is difficult to get from food because there isn't a lot of it and what chromium exists is often not absorbed and used by the body.[21] The mere presence of chromium in food is no guarantee that health benefits can be derived from it. Chromium's positive charge must be neutralized before it can approach and eventually enter living cells. Thus, chromium has to be attached to special chemical substances, called chelators, that deactivate the mineral. Yeast and kidney extracts worked well because they contain good chelators.

All inner surfaces of the body, including the intestines and the blood capillaries, are lined with cells. All cells, in turn, have a protective fence or wall, called the membrane, around them. Cell membranes are like the walls and fences around high-security areas—visitors don't just come and go as they please. The barrier that surrounds cells is made of specialized chemicals designed to reject many of the substances coming into contact with it. Cells need this protective barrier to keep from being flooded with water or overcome by too much of any one substance in the diet.

Chromium ions, like all mineral ions, have to be tucked inside the chemical chambers of chelators so they can be smuggled through membranes to get them from the intestines to the bloodstream. Schwarz and Mertz recognized this fact when they first discovered that chromium is an essential mineral. The two scientists knew chromium was only one component of what they called the GTF because their early studies proved chromium salts were not as effective as chelated chromium in improving insulin function.

Dr. Mertz and other investigators spent a good deal of time searching for the natural or, at least an ideal, chromium chelator, the substance that made yeast and kidney extracts effective. They were searching for a chelator capable of effectively masking the charge of the chromium ion so this important mineral could be smuggled inside human and animal cells. Had such a chromium chelator been discovered earlier, chromium would have readily been accepted without question as essential for humans, and the incidence of heart disease, obesity, and diabetes would be less than what it is today.

This chemical, named picolinate, was discovered in living cells in the late 1950s,[22] but was not fully appreciated at that time. My colleagues and I at the USDA rediscovered picolinate as part of another research effort—the search for a chemical we suspected was missing in children who became very sick and often died because they couldn't absorb zinc.

Clues From A Children's Disease

In the mid-1970s, an English pediatrician discovered that an inherited disease called acrodermatitis enteropathica (AE) could be cured by simply feeding patients about five times the recommended daily requirement of zinc.[23] AE, a genetically transmitted disease that affects children, causes a severe rash, loss of immunity to disease, and drastic behavioral changes. Untreated, it can lead to death.

This discovery led us at the USDA to suspect that AE develops because these children cannot get along on the amount of zinc normally present in the diet, since the children were not eating diets deficient in zinc. This made me suspect that AE occurred because the children could not make a zinc chelator needed to move zinc from the intestines into the bloodstream.

After reading the papers of others, and doing some work of our own, we decided that there were at least three noteworthy points about this disease. These observations directed the course of our research:[24]

1. The symptoms of AE start after the child has been weaned from mother's milk, and human milk had cured many children.

2. A drug called hydroxyquinoline had been successfully used to treat AE, but nobody knew exactly why it worked.

3. In many of the research reports, children with AE had chemicals in their urine not found in the urine of normal children. These chemicals were formed in the process of breaking down tryptophan, an amino acid.

We learned that the drug hydroxyquinoline is a good chelator for zinc. From this, we decided that the drug was effective in stopping AE's symptoms because it could easily combine with zinc in the intestines and act as a substitute for a naturally produced chelator. Therefore, since human milk also cured AE, we decided to start looking for a zinc chelator in human milk.[25]

We also looked at the possibility that something was wrong with the handling of tryptophan in these children's bodies. Tryptophan, like all amino acids, must be dismantled in an orderly, step-by-step process so the cell can make use of the individual chemicals. If anything goes wrong on the disassembly line, the chemicals that were made before the blocked step start to accumulate. They build up to a point where they eventually spill out of the cell and end up in the urine. This is why doctors were finding tryptophan break-down chemicals in the urine of AE patients.

In reading about tryptophan, I found a chemical that was created near the end of the tryptophan disassembly line: picolinate. I saw it was an ideal metal ion chelator, and that some of the chemicals that are formed before the picolinate stage are the same chemicals found in the urine of children with AE. We tested human milk, infant formula, and cow's milk, and discovered that only human milk contains measurable quantities of picolinate. When we tested picolinate, we realized we had at last identified one of nature's naturally produced metal ion chelators.[26-28] It was just the thing we had been looking for.

The Detroit Connection Proves Invaluable

Shortly after we identified picolinate in human milk, I received a telephone call that provided a vital connection in proving picolinate's importance in the body.

Dr. Inga Krieger at Detroit's Children's Hospital called to tell me about an infant girl who had developed the symptoms of AE at four months. Dr. Krieger had started giving the little girl a commercial pancreatic extract, and noted that the symptoms of AE disappeared. Later, the girl was treated with daily supplements of oral zinc, which were effective only when the daily dose exceeded 60 milligrams zinc—several times the amount needed by a child that age. Dr. Krieger contacted me after tests showed that the pancreatic extract contained less than 5 milligrams of zinc! The pancreatic extract, like human milk, was effective because it contained picolinate.[29]

Dr. Krieger then started giving her young patient daily zinc picolinate supplements. The girl was given only 5 milligrams of zinc a day, considerably less than the 60 milligrams a day being used to treat her at the time our study began. The difference was that the 5-milligram dose was in the form of zinc picolinate, compared with the 60-milligram dose in the form of zinc salts. The girl remained symptom-free during the entire time she was treated with zinc picolinate.

Dr. Krieger then enlisted the services of Dr. Ralph Cash, who was treating AE patients on a regular basis. They obtained parental permission to discontinue treatment in about a dozen cases until the rash and other symptoms of AE reappeared. They then started treating the patients with low doses of zinc picolinate. Again, the symptoms disappeared.

The medical observations from this study were very exciting and encouraging, especially when we measured picolinate in the blood of both normal children and children with AE. We discovered that the amount of picolinate in the blood of normal children was three to four times greater.[30]

The observations with AE children provided some of the first sound proof that humans cannot absorb mineral ions without the aid of chelators. Children who inherit AE must be treated daily with large amounts of zinc unless they are treated with zinc picolinate. These clinical studies prove that the human body must produce picolinate to extract zinc from the diet and move it into the bloodstream. When too little picolinate is produced, zinc is not extracted, and the body starts to malfunction for lack of this essential nutrient.

These studies completely changed my approach to nutrition. I no longer considered it possible for metal ions to pass unaided from the intestines into the bloodstream. The mere presence of a mineral ion in a food was no guarantee the nutrient would be absorbed and utilized by the body—ions by themselves don't provide necessary health benefits. Eventually, I and many others proved that chromium chelated with picolinate is a health-giving nutrient.

CHROMIUM PICOLINATE IS THE SOLUTION FOR CHROMIUM SHORTAGE

Picolinate is an ideal metal ion chelator, but our studies show that most humans barely make enough picolinate to chelate all the zinc ions needed to maintain optimum health. The amounts of copper, iron, manganese, and zinc in the average diet are all several times the amount of chromium. This explains a major problem with chromium and health: not only is chromium scarce in food, but what chromium exists is hardly used because the body's primary metal ion chelator, picolinate, is tied up by other ions.

That is why chromium picolinate supplements are necessary for health. Chromium is needed to keep insulin working properly, but what chromium we eat is difficult to absorb. Picolinate is an ideal chelator for chromium and other ions because it is naturally produced in cells and readily combines with metal ions. Although picolinate can be made by the body's cells, not enough is produced to hunt down and attach to the tiny amounts of chromium in the diet. Natural sources of chromium picolinate include brewer's yeast, liver, and kidney. Unless an individual eats a lot of these foods every day, chromium is nearly impossible to get from the diet without daily supplements.

5. CHROMIUM AND INSULIN—A HEALTHY CONNECTION

All of the benefits seen in chromium picolinate studies come from improved insulin function, which results from increased amounts of chromium reaching the cells where it is needed.[1-13] Body cells need chromium to keep insulin working properly. Insulin directs the movement of digested food into the proper cells and affects how that food is used. It is usually thought of in connection with blood sugar or diabetes. The fact is, however, that insulin regulates many other processes in the body. When it doesn't work, many bodily functions are adversely affected.

In this chapter, we'll first look at the chromium shortage, and how that results in poor insulin function. We'll then look at exactly how insulin works, and how chromium helps insulin to do its job.

AN EPIDEMIC: THE LACK OF CHROMIUM IN OUR FOOD

There's a good explanation for the success of chromium picolinate experiments. To put it bluntly, we have an epi-

demic in this country. It's not an epidemic caused by any
bacteria or virus. The cure for this epidemic isn't any drug.
The cure is chromium picolinate, because this epidemic is
caused by a lack of chromium in our diet. Without ample
chromium in the body, blood sugar and fat cannot be properly regulated. This results in food cravings, followed eventually by obesity, heart disease, and diabetes.

Because of its central role in helping control the way digested food is used, chromium is one of the most important
minerals in the body. Unfortunately, it is also one of the most
difficult to obtain from our diet. There is very little in food,
and what chromium does exist in food is difficult for the body
to assimilate and use.

Everything we know about chromium as a nutrient is tied
to the action of insulin, and malfunctioning insulin is at the
root of the epidemic caused by the lack of chromium. Insulin
impotence is a very appropriate name for the problem. Impotence is defined as the inability to act. There isn't a better
way to describe the course of this disease than to call it the
inability of insulin to act. It could also be appropriately
named pre-diabetes, but that implies everyone who gets the
disease will end up with diabetes, which is not always true.
As is the case with all epidemic diseases, some people will
suffer greatly while others will have only mild symptoms.
Some will have difficulty developing all the muscle they
want while others will have difficulty controlling the fat on
their bodies. The much less fortunate will develop heart
disease. Others will become obese and diabetic.[14-16] Whether
the symptoms are mild or extensive, health is affected.

People with insulin impotence don't have problems producing insulin, they have problems using the insulin they
produce. Since insulin is not used efficiently, blood-sugar
levels become too high, which causes problems. Eventually,
most people with insulin impotence end up making too
much insulin, which only adds to their problems. I suspect

a lot of people have insulin impotence and don't even know it. The medical problems associated with this disease go undetected for years. The condition is what doctors call an insidious disease—it's silent and deadly. Everybody who eats is exposed to it.

Insulin impotence and the chromium deficiency that plays a part in causing it are aggravated by the typical American diet. Most of the food we eat has a lot of sucrose, which is a combination of glucose and another sugar called fructose. Sucrose is sweet, and so is it added to food to make it taste better. It is the main ingredient in candy, cookies, cake, and other sweets. Estimates indicate that between 20 and 40 percent of the calories we take in are from sucrose. When I was young, adults told kids not to eat too much candy because it would cause "sugar diabetes." This is an example of an old wives' tale that turned out to be right on the money.

Also, we are encouraged to eat diets low in fat and high in carbohydrates. Carbohydrates contain mostly glucose, a contributing cause of insulin impotence. Several studies and books[17] address the unhealthy practice of consuming too much carbohydrate. You should get about a third or less of your daily calories from fats, and you should make up the extra calories with a balance of protein and such complex carbohydrates as pasta and potatoes. Complex carbohydrates are digested more slowly, and thus release glucose into the bloodstream more slowly.

INSULIN IMPOTENCE STARTS EARLY

Insulin impotence starts soon after a person is born. Many preschool children probably have the problem. The very early phases of the disease are caused primarily by lack of chromium in the body. Later in life, the disease becomes more complicated because, in addition to a lack of chromium, there's also too much sugar in the diet.

The studies done by Dr. Henry Schroeder, whom we met in Chapters 1 and 4, show that chromium is depleted as humans grow older.[18] All the experiments with pigs and rats in Chapter 3 show that chromium is less than adequate soon after birth. During the early phases of insulin impotence, insulin is not totally effective because it can't get where it's supposed to go and do what it's supposed to do efficiently. As we grow older and eat more food laced with sugar, insulin's effectiveness erodes even more because the body's cells have to protect themselves from being overcome by too much glucose. Too much glucose causes glycation, in which glucose attaches to cell parts. Since glycation can be very harmful to a cell, the cell responds by not letting the glucose inside the cell membrane.

This failure to respond to glucose triggers a chain reaction within the body. The muscles temporarily switch off their insulin recognition system, causing glucose to be diverted to fat cells and to the liver.[19] If the sugar onslaught continues, fat cells eventually shut off their insulin recognition system. This is followed by an almost total shutting down of the glucose transport system—the system that gets glucose from the outside of the cell to the inside—in muscle cells and later in fat cells.[20] Eventually, glucose spills over into the urine and often triggers processes that stop insulin production in the pancreas. At that critical stage, the disease has progressed from insulin impotence to diabetes. (See "How Healthy Is Your Insulin?" on page 65 for information on a test that would let you know how well your insulin is working and Chapter 7 for more information on diabetes.)

As strange as it seems, too much glucose can also lead to too much insulin. As insulin becomes less and less effective, glucose remains in the bloodstream much longer than it should. The presence of too much glucose in the bloodstream causes the pancreas to produce and secrete more insulin. This

How Healthy Is Your Insulin?

There is a way to tell if your insulin is not working properly before diabetes develops. It is called a glucose tolerance test. Since this test must be performed in a clinic or medical laboratory, you will have to speak to your doctor first. You may want to take this test if you have a strong family history of diabetes, or if you have had problems with mood swings or food cravings, or if you have such standard diabetes symptoms as unusual thirst, frequent urination, unexplained weight loss, or drowsiness.

Before taking the test, you will be asked to not eat anything for twelve hours. This will ensure that the test results are not thrown off by glucose already present in your system. Then, a small sample of blood will be taken from your fingertip. This will provide a baseline for the laboratory technician to measure your glucose levels against.

Next, you will be given a known amount of glucose, generally in the form of a very sweet drink. After that, blood will be taken every half-hour over a period of a few hours. This will allow the technician to see how quickly your body is able to remove glucose from your bloodstream. You may also be required to give a urine sample, which will allow the technician to see if any glucose was spilled into the urine.

If this test indicates that you have diabetes, see Chapter 7 and speak to your doctor before using chromium picolinate.

causes high blood-insulin levels, known as hyperinsulinemia. High levels of sugar and insulin in the bloodstream make for a very unhealthy combination. Insulin is a vital hormone, but too much is as bad as too little, and can result in problems such as obesity and increased chances of heart disease and diabetes. High blood-insulin levels also cause premature aging, and more likely than not contribute to osteoporosis (see Chapter 9).

HOW DID WE COME TO DEPEND ON INSULIN?

By this point, you might wonder how we developed a system that depends on such a seemingly delicate balance. The reason is that the human body is adapted to deal with a scarce food supply. The abundance—or overabundance—of food we enjoy in modern life is a very recent development.

Millions of years ago, when Nature was experimenting with the design of living things, one lesson was learned right away. Most creatures, particularly humans, eat intermittently rather than continuously. To keep enough glucose in the blood to fuel the nerves, which require a steady supply of it, something was needed to direct glucose and amino acids into the muscles during a meal, when the supply is plentiful. The design that evolved uses glucose from food to trigger release of a chemical—insulin—capable of opening glucose gates in the muscles.

When you eat a meal, the food is broken down into substances that can be dissolved in bodily fluids, or at least carried on the back of a protein that dissolves in fluid. For example, a piece of bread is made of the carbohydrate starch, which is hundreds of glucose molecules hooked together. Saliva and the juices in the intestines contain enzymes that break the starch apart. What starts out as a long chain ends up as hundreds of individual glucose pieces. Glucose can easily be dissolved in fluid and is quickly transferred from the intestines to the bloodstream.

The first organ the glucose passes through after coming from the intestines is the pancreas, a long gland that sits behind the stomach. The pancreas contains specialized cells that produce insulin. The flow of glucose through the cells of the pancreas causes the release of insulin into the bloodstream. Insulin then travels around the body, informing certain cells that there is an excess supply of fuel in the blood. Responding to the message from insulin, cells open the gates that allow glucose to enter.

A CLOSE-UP VIEW OF INSULIN IN ACTION

If you're a science fiction buff or watch the late shows, you may have seen a movie from the mid-1960s called *Fantastic Voyage*, an Isaac Asimov story. In it, the main characters are put into a submarine, miniaturized, and injected into the bloodstream of a man whose life they have to save.

In my version, there would be a scene where the ship docks at a muscle or fat cell for routine maintenance. Without warning, the fluid around the cell becomes murky and filled with chemical debris—the host has just finished digesting a meal.

Soon a floating object appears beside the cell. It looks like a type of sailboat called a catamaran, with two long amino-acid hulls hooked together side by side. It is insulin, come to clear the debris from the water. The insulin floats up and attaches to a protein dock on the surface of the muscle cell. From their vantage point, the crew can see that the receptor dock has mooring posts imbedded in the cell membrane. The dock extends all the way through the membrane to the inside of the cell.

When the insulin bumps into the dock, the shock waves reverberate all the way to the inside, where a chemical messenger is asleep. The shaking awakens the messenger, who runs to the glucose transport depot to announce the arrival of the insulin ship. Knowing that the arrival of insulin is always accompanied by a gaggle of glucose molecules anxious to get inside, drivers of the glucose transport trucks haul protein tunnels to the membrane. The tunnels are pushed through to the surface, where hundreds of glucose molecules rush through the openings.

On the surface of the cell, our crew watches a curious scene. Soon after the insulin bumps into the receptor protein, the whole dock, with the insulin ship attached, starts moving slowly along the surface of the cell membrane. The whole entangled protein gob soon disappears into a pit or cave on the surface. As soon as the dock disappears from view, a new one, without any insulin attached, appears on the surface. Another insulin ship floats up and the whole scene is repeated.[21]

What does chromium have to do with any of this? And how does the system go awry?

Chromium and a Slimmer, Trimmer Insulin

As we've seen, insulin must find and attach to a receptor before a signal can be sent to the cell's interior. To get to a cell, insulin has to travel through the bloodstream until it reaches a tiny blood vessel, known as a capillary, next to its destination. It must then be able to pass through the walls of the capillary. The more streamlined the insulin, the easier the passage through the capillary walls.

Once it reaches the fluid in the harbor around the cell, the insulin has to effortlessly move into its designated dock. If it's too big, it will scrape the sides of the protein pier, which makes mooring difficult. If it's too small, it might get tossed around in the bay, which also makes mooring difficult.

Insulin must be a certain size and shape so it can slip through the capillary and slide into port. When insulin can't move from the blood or can't dock easily, it can't stimulate glucose clearance. While we are not yet sure of exactly what chromium does, it could play a role in helping insulin keep just the right shape so that the insulin molecule can maneuver through capillary walls and fit into the cell's ports.

Scientists have long known that metal ions cause proteins to be pulled and twisted into specific shapes, usually more streamlined shapes. Insulin isn't a particularly large or bulky protein, but like most, it can stand some streamlining. Chromium picolinate seems to attach to insulin, so there's a good possibility that chromium helps insulin by keeping it in the correct shape. Researchers have observed much faster clearance of insulin from the blood of animals fed chromium picolinate supplements—in several experiments with pigs, insulin was removed from the bloodstream 40 percent faster.[22,23] A trimmer, sleeker insulin, shaped by chromium, might be the explanation.

Chromium, like other ions used in the body, acts as a cofactor. Cofactors are substances needed by enzymes and other proteins to perform their functions. Copper and iron ions are needed by many enzymes to help attract and hold the chemicals they assemble or break down. Zinc ions are needed by some proteins for the proteins to maintain the correct shape. Many proteins are useless without an accompanying metal ion cofactor, and insulin is a protein. My students and I, along with other scientists, continue to gather evidence that chromium picolinate binds directly to insulin.

One scientist thinks that chelated chromium—chromium in a form acceptable to living cells, such as chromium picolinate—attaches to both insulin and to the cell receptor, forming a bridge.[24] That is a definite possibility. It is also conceivable that chromium picolinate combines with the insulin itself, pulling it into a shape more readily recognized by the receptor. The chromium picolinate might also move inside the cell, where it works with some enzyme or protein needed to take glucose inside the cell.[25,26]

Insulin communicates with cells through receptors on the cell membrane. Only one call is allowed through, so once insulin combines with a receptor, that communication line goes dead. Before a receptor can be used again, it must first get rid of the attached insulin. Receptors carry the insulin inside the cell, where it is detached. The receptor then returns to its location on the outside of the cell membrane. One experiment showed that chromium picolinate improves the efficiency with which cells remove insulin and return receptors to the surface.[27]

There is ample evidence that chelated chromium improves the efficiency of insulin action. Whether chromium interacts with insulin directly, or with insulin's target cells, or with both, insulin is much more effective when chromium is present. Because insulin needs chromium for optimal action, chromium is a cofactor for insulin. I prefer to call it insulin's chrofactor.

Glucose Alone Can Gum Up The Works

Ineffective insulin can be linked to diets that contain too much carbohydrate, especially simple sugars. Insulin impotence starts out as a harmless attempt by muscle cells to prevent glucose from accumulating.[28]

Glucose inside a cell is never allowed to simply float around because it can combine with proteins and enzymes, interfering with their operations. Glucose is used primarily to produce energy. If the cell doesn't need energy the glucose is then diverted to other areas, where it is used as raw material to make chemicals the cell does need. Excess glucose is stored away as glycogen, which consists of long chains of glucose molecules. Glycogen is big and takes up a lot of space. Space is limited inside the cell, so there is a set amount of space for glycogen. Once glycogen has filled that space, no more is made.

Cells have devised an ingenious method to keep too much glucose from getting inside and gumming up operations.[29] When there is more glucose than the cell needs for energy, raw material, or storage, the excess glucose interferes with the communication pathway leading from the insulin receptor to the glucose transporters. It does this by ambushing the messenger who is supposed to notify the glucose transport depot that insulin has arrived. When no message is transferred to the glucose transporters, there is no movement of these proteins to the membrane. When there are no glucose transporters at the membrane, no glucose moves inside the cell. This process is efficient because it allows the cell to protect itself from too much glucose, but can easily be reversed.

Ambushed Messengers and Unused Transports

Ambushing the chemical messenger inside the cell is intended as a temporary means of protecting the muscle cell from accumulating too much glucose. During this protective action, glucose is diverted to the fat and liver cells. As soon as the muscles use up the excess glucose, the messenger is set free. Glucose can again enter the muscle cells through the protein tunnels.

However, if high levels of glucose continue to surge through the bloodstream, fat cells eventually become threatened by too much glucose. Like the muscle cells, the fat cells are then forced to take protective action by limiting the amount of glucose that is allowed inside.[30] This is also intended to be a temporary measure.

These temporary measures are intended to protect cells until a glut of glucose no longer threatens the cells. Too often, because of poor diet and a lack of exercise, large amounts of glucose continue to circulate in the bloodstream. The body's cells react to the continued high levels of glucose by no longer making glucose transporters. After all, the storage depot, where the protein tunnels are kept, already has a supply of tunnels that aren't being used. Cells are not wasteful, so glucose transporter production is stopped, first in the muscle cells and later in the fat cells.

When muscle and fat cells ignore insulin's call for glucose

transporters, they do so to prevent too much glucose from getting inside and fouling up normal operations. However, this leaves the glucose nowhere to go, so it just stays in the bloodstream, where it can eventually lead to the overproduction of insulin.

CHROMIUM GETS KICKED AROUND A LOT

Unfortunately, the cells' attempt to protect themselves makes insulin's job even more difficult. Scientists with the USDA discovered that diets high in carbohydrates—especially simple sugars—cause people to excrete more chromium through the urine.[31] Thus, too much carbohydrate keeps cells from responding to insulin, and at the same time causes loss of the chromium that is needed to keep insulin functioning efficiently.

Once inside the body, chromium is assaulted by a whole host of attackers. Not only do sugar-rich diets cause chromium loss, but so do exercise and physical trauma. Very strenuous exercise, such as running a few miles at top speed, causes a big increase in the amount of chromium excreted in the urine.[32] Athletes who work out regularly at the maximum pace are prime candidates for developing insulin impotence.

Moderate exercise doesn't seem to cause chromium losses. But keep in mind that insulin malfunction is a silent, slow-acting attacker. All of the athletes I've worked with have benefitted by taking chromium picolinate supplements—a subtle admonition that fit is not always healthy.

The stress of strenuous exercise is probably what causes chromium to be lost, because physical trauma does the same thing. People being treated at an emergency shock treatment center excreted very high amounts of chromium in their urine.[33] Without a doubt, if levels of urinary chromium were studied in more diseases, we would have a long list of bodily stresses that lead to chromium loss.

It is chromium loss, when combined with the body's inability to use the chromium found in food, that leads to poor insulin function. Poor insulin function and low chromium levels together lead to a number of physical ailments, such as obesity (Chapter 6), diabetes (Chapter 7), and the various consequences of aging (Chapter 9). Low chromium and poor insulin function also lead to an inability to develop muscle (Chapter 8).

6. CHROMIUM PICOLINATE AND FAT LOSS

I n the early 1990s, a lot of press was generated when Oprah Winfrey lost sixty-seven pounds. A year later, millions of Americans empathized with Oprah after learning she had regained eighty-four pounds.

Every year, millions of people spend billions of dollars on weight-loss plans and programs. Celebrities promote videotapes, and the best-seller list always has a new "fat burner" book. But despite valiant efforts, many dieters find that the fat comes back.

In this chapter, we'll discuss why diets that don't include adequate chromium don't work. We'll see how insulin distributes glucose and fat, and why the fat stays when chromium supplies are inadequate and insulin malfunctions. We'll also look at how energy, fat, and temperature are interrelated. Finally, we'll come up with some recommendations for using chromium picolinate as part of a sensible fat-loss program.

NO CHROMIUM, NO FAT LOSS

Most reduction plans involve a combination of both exercise

and a low-calorie diet—or worse, appetite suppressants—
that drastically cuts the supply of fuel to the body. On paper,
the theory is plausible. Burn more calories than you take in
and you will deplete the fat cells of fuel.

The theory is great, but it just doesn't work. Over 90
percent of fat-loss programs fail simply because the food
control systems of the body aren't working properly. Why
isn't fat loss a simple matter? The answer lies in the nature of
the body's fuel-supply system.

The main fuel burners in the body are the nerve cells and
the muscle cells. Nerve cells are found in the brain, spinal
cord, and nerves. The brain regulates body functions by
constantly communicating with other organs through the
spinal cord and the nerves. Therefore, nerve cells require a
constant source of energy. Glucose is the only fuel these cells
will use to make energy, except under starvation conditions.

Muscles are needed for movement. Skeletal muscles, like
those in our arms and legs, are attached to ligaments hooked
to bones. When the muscle contracts, the bone moves. A lot
of energy is needed in the process, especially during exercise.
Unlike nerves, muscles can and do use both glucose and fat
for fuel, but prefer fat because of the higher amount of energy
fat provides. Ounce for ounce, there is far more energy in fat
than in glucose. Muscles don't store fat, which is kept in fat
cells.

Nerve cells don't store glucose, so they need a continuous
supply from the blood. This wouldn't be of any significance
if it weren't for the fact that muscle cells burn far more fuel
than nerve cells. Fortunately, muscle and liver cells store
glucose for the times when there is a great demand. This
prevents the nerve cells from dying for lack of fuel.

The problem for muscle cells lies in getting the fat out of
the storage areas quickly enough to be used as fuel. Insulin
controls that process. If insulin is not operating properly, it
is very difficult to move fat from the fat cells to the muscle

cells. If fat can't be used as a source of fuel, the muscle cells use stored glucose and amino acids for fuel. When these are used up, the cells' operations slow down. The cells may eventually stop functioning and waste away. The end result is a loss of muscle.

At first, this helps the dieter lose weight because muscle is much heavier than fat. This gives the dieter a false sense of achievement. But the lost muscle cells are replaced by special cells whose purpose is to putty in blank spaces. Eventually, those cells turn into fat cells, and the beleaguered dieter ends up with more fat than at the start of the program. That is what has come to be known as yo-yo dieting.

Insulin is one of the most important fat regulators, and if chromium intake is not adequate, insulin doesn't function properly. Therefore, without a sufficient supply of chromium in the body, there's no hope of shedding excess fat and keeping it off. Controlling body composition—the balance between fat and muscle in the body—is a real problem in this country because less than 10 percent of Americans get enough chromium in their diets. And, as we've seen, much of the chromium that is in food isn't in a useable form.

That's why we need a useable, or bioavailable, form of chromium—like chromium picolinate. The human and animal studies discussed in Chapters 2 and 3 prove that chromium picolinate supplements can help the body to both lose fat and prevent fat accumulation, and to do so without losing muscle.

INSULIN, THE FAT DIRECTOR

Insulin controls the distribution and use of both glucose and fat. That is why insulin is the master fuel-supply hormone. It's also why insulin controls how much fat the body accumulates.

From Glucose to Fat

Insulin causes glucose to be transferred into muscle cells, liver cells, and fat cells, in that order—I call it the glucose pecking order. When muscles are used frequently, as in regular exercise, they burn a lot of fuel to make energy. In the process, stored glucose gets used and needs to be replaced. To replace the spent fuel, much of the glucose from food will be moved into muscle cells via insulin. Nerve cells don't need insulin because glucose can move into them freely. Nevertheless, the amount of glucose in the blood has to be kept at a level sufficient to keep the nerves from slowing down and stopping.

The amount of glucose is kept fairly constant with the help of the liver, which stores glucose to be dumped into the bloodstream if the level drops too much. If a person exercises often, the nerves, muscles, and liver together should use all the glucose made available after every meal.

But if a person doesn't exercise regularly, the muscles don't need to replace their fuel supplies. Under these conditions, not much glucose is used. The excess is picked up by the fat cells, which convert the glucose into fat. The same thing happens if a person overeats. Again, there is excess glucose in the bloodstream that has to be stored away.

From Fat to Fat

The fat you get from food doesn't go into the bloodstream by itself. That's because fat can't be dissolved in liquid. As fat moves through the intestines, it is attached to proteins, which can dissolve in the bloodstream. Laden with fat and cholesterol, these proteins move through the bloodstream. They ferry fat from cell to cell, depositing it where it's needed.

Because the fat is attached to protein, it must be released before it can move into cells. Specialized enzymes are needed to untie the fat from the protein. Those enzymes, which are attached to the walls of blood vessels, are activated by insulin. When the blood-insulin level goes up, the untying enzymes get turned on and release fat from the carrier proteins. Since insulin is already in the bloodstream during digestion to help control the use of glucose, it makes a lot of sense for insulin to help direct fat storage as well.

As digestion progresses, the blood-insulin level is supposed to decrease. That allows some of the fat to stay in the bloodstream, readily available as needed. But when a person is affected by hyperinsulinemia, that doesn't happen. (See Chapter 5 for information on how this condition develops.) Hyperinsulinemia is characterized by high levels of insulin—called renegade insulin—in the blood. The renegade insulin keeps the untying enzymes activated, and they just keep sawing away at the fat. Most of the fat is taken up by fat cells because muscle cells only take in fat when they need it for energy.

WHEN INSULIN MALFUNCTIONS, THE FAT STAYS

There are other reasons people become obese when insulin can't do its job. Part of the problem comes from the way glucose and fat are processed in the muscle and fat cells. Another part of the problem comes from the way glucose and insulin affects the brain. When too much glucose and too much insulin flood this complex, interrelated system, the body accumulates more fat than it can get rid of.

Insulin Problems in the Body

You may have heard of triglycerides. These are simply three strings of fat hooked to a chemical named glycerol. Fat cells store fat in the form of triglycerides, and they are absolutely useless for anything except storage.

The muscles prefer fat as a source of energy, but since

triglycerides cannot be sent directly to the muscles, the fats have to be unhooked one at a time from the glycerol so they can be attached to proteins and carried to the muscle cells. Fats are detached by enzymes that are turned on only when the blood-glucose level goes down. Since this level doesn't go down when cells don't respond to insulin, these enzymes don't see much action, which means that the triglycerides stay in the fat cells.

As you can see, insulin impotence makes it very difficult to control body composition. The enzymes used to release fat from blood proteins for movement into fat cells are switched on by insulin, and the enzymes which detach fat from glycerol so it can move into the blood are switched off when insulin isn't functioning. Thus, fat moves into fat cells easily, but gets trapped there and can't leave.

Insulin impotence also influences fat production through its effect on a hormone called DHEA. DHEA keeps glucose that enters a fat cell from turning into fat.[1] But high levels of insulin in the blood inhibit the body's ability to produce DHEA.[2] Thus, too much insulin leads to obesity.

Fat loss would never become a concern if fat didn't accumulate in the body. A British doctor named F. W. Fox has studied the puzzle of obesity. According to Dr. Fox, body weight is either stable or increases slowly during adult life.[3] Any upward drift amounts to only a half-pound a year between the ages of twenty and sixty. If this were true, a forty-year-old individual should have gained, at the most, only ten pounds. This is hardly enough to make over 33 percent of Americans drastically obese. But they are, and it's probably because they don't have enough chromium in their bodies.

Insulin Problems in the Brain

From the moment insulin becomes misshapen from lack of chromium, control of body composition becomes a problem.

Cravings send people to the cookie jar or candy machine, where they load up on the fuel they think they need. In fact, their fuel tank is probably already full, but the fuel gauge isn't working.

Our bodies have a built-in fuel gauge designed to signal when it's time to eat. That fuel gauge is a specialized batch of cells in the brain called the satiety center. These cells regulate body composition by controlling hunger, stopping insulin secretion, and causing excess calories to be burned.[4-14] The satiety center cells contain monitoring devices that measure how much glucose is getting into the cells. Since the monitoring devices are inside of the cells, the amount of glucose in the bloodstream is not measured correctly when insulin isn't working as it should.

Like muscle and fat cells, satiety cells take in glucose when insulin attaches to receptors on the cells' surfaces (see Chapter 5). By measuring the amount of glucose in these cells after insulin attaches to them, the satiety center takes stock of both glucose and insulin availability.

The satiety center is in the area of the brain called the hypothalamus, one of the body's command centers. Because hypothalamus cells are nerve cells, they are connected to other cells that start or stop bodily operations as needed. As long as glucose enters these cells, the command center detects an ample fuel supply, so that the desire to eat is blocked.

The desire to eat comes from brain cells communicating with the satiety center. When the amount of glucose drops, a low-fuel state is detected and communicated to areas of the brain capable of creating the sensation of hunger.

Like muscle and fat cells, satiety center cells must have insulin attached to them before they can take in glucose. Insulin impotence causes a false low-fuel reading by the fuel gauge because glucose doesn't get in to turn off the desire to eat. I believe that this happens when insulin can't reach the satiety cells. These cells, like the rest of the hypothalamus, are surrounded by capillaries with tight walls. As we saw in Chapter 5, insulin must be able to get through these walls before it can reach the cells. Misshapen insulin can't do that.

If the satiety center cells don't take an accurate measure of the amounts of glucose and insulin, they respond as if the amount of available fuel is low. A nearly constant hunger sensation is created. There is also no regulation of insulin secretion, which means that even more excess insulin is produced.

The end result? When you're hungry, you eat. The more often you eat, the more food you take in. If you don't burn the fuel in the food you eat, you become obese.

ENERGY, FAT, AND TEMPERATURE

There are different ways the body uses energy, some of which you are not consciously aware. For example, whether you are sitting on a couch or moving a piano, your cells need some fuel to produce energy. Some of the energy is used to maintain body temperature, some is used to assemble or dismantle chemicals and, in the case of the nerves, some energy is used to send messages to other cells.

Regardless of how the energy is used, fuel has to be burned. Just like any fire, that requires oxygen. Oxygen is not stored inside cells, so it has to be brought in continuously. Since oxygen comes from the air, the lungs have to constantly expand and contract to bring in the air. Expansion and contraction is accomplished by muscles that use a lot of fuel to do their job. Once the oxygen is in the blood, it has to be moved from the lungs to cells throughout the body. Blood is moved by the constant working of the heart muscle. That also requires a lot of fuel.

These unconscious processes are what keeps the body alive. We call this the basal metabolic rate (BMR), and it accounts for about two-thirds of all the fuel used in a day.[15]

The amount of energy used to sustain the BMR is surprisingly high. A average-sized man burns about 1,650 calories a day just to keep his body functioning. An average-sized woman uses about 1,500 calories per day.

Since the BMR uses the major portion of the calories consumed each day, a reduction in this activity leads to a major cutback in the rate at which calories are burned. The BMR is regulated in large part by a hormone called triiodothyronine (T3). Diabetes causes a decrease in the production of T3 but proper treatment with insulin restores production of the hormone to near-normal levels.[16] These observations prove that the production of T3 is affected by either a lack of insulin or the poor action of insulin. It also proves how important insulin is in controlling the BMR.

Body cells have to maintain a constant temperature or they won't work properly. An increase in temperature is detected by brain cells which, in turn, cause the sweat glands to secrete water onto the surface of the skin where it will evaporate, causing cooling.

Muscle and a special kind of fat tissue called brown adipose tissue keep the body warm. When a drop in body temperature is detected by the nerve cells, an alarm is sent to the brain. The brain then triggers the release of chemicals prompting the body to burn some stored fat. The burning of fat produces heat.

In addition to keeping us warm, muscle and brown adipose tissue are supposed to burn up any excess fuel we take in. The same brain cells that monitor glucose levels—the satiety center cells—also communicate with the cells built to get rid of excess fuel. If more than a certain amount of glucose flows into the satiety center cells after insulin attaches to them, an alarm is sounded. Neurons in the brain send a message to the brown adipose tissue and muscle, causing fat to be burned. This process is known as thermogenesis, the creation of heat. Thermogenesis is another method the body uses to keep excess fuel—fat—from being stored in the fat cells.[17–19] Depending upon what and how much you eat, this very important body function uses between 200 and 300 calories a day.

Insulin impotence has a big effect on fat storage because it essentially disables thermogenesis. Glucose can't get inside the satiety center cells until insulin attaches to them. If insulin can't make its way to the attachment sites or can't attach effectively, or if the cells simply don't respond, glucose doesn't flow into the cells. Since the monitoring cells incorrectly detect a low-fuel situation, they don't send any instructions to use stored fuel. Fat, which would normally be burned as excess, accumulates.

To make things worse, the body has a more difficult time maintaining a constant temperature when the cells stop burning fat. More insulation is needed, leading to production of more fat cells to provide the extra blanket of insulation.

According to some investigators,[20] malfunctioning insulin causes the retention of about 125 calories a day. That amounts to about ten pounds of fat a year. That extra fat, along with the other ways in which insulin malfunction hinders fat-burning, make it very difficult to control body composition. Since chromium picolinate helps keep insulin functioning effectively, it helps insulin to control body composition.

USING CHROMIUM PICOLINATE TO HELP LOSE FAT

As we saw at the beginning of this chapter, cutting food intake to lose fat—unless you are genuinely overeating—is not only unnecessary but foolish, because it leads to muscle breakdown. Dr. Gil Kaats and his associates in San Antonio, whom we met in Chapter 2, have devised a plan that should promote fat loss and prevent muscle loss.[21] It's a plan you might want to consider trying. Chromium picolinate plays a vital role in any program for fat reduction because it keeps insulin working properly.[22,23]

One of Dr. Kaats's studies showed that a person can lose twelve pounds of body fat in only two months—without exercise—by taking the following steps:

- Increasing the fiber content of his or her diet by eating more fruits, vegetables, and grains.
- Moderately restricting calorie intake to about 2,000 calories for women and about 2,500 for men. This seems like the hardest step, but the chromium picolinate and the fiber will help prevent food cravings.
- Taking 200 milligrams of carnitine—an amino acid that helps clear fat from the body—and between 400 and 600 micrograms of chromium picolinate a day.

With moderate exercise, one can achieve the same fat loss in ten weeks by eating a diet that provides less than 30 percent of its calories from fat and by taking between 400 and 600 micrograms of chromium picolinate a day, with no carnitine. These goals can be accomplished without losing either muscle mass or energy, problems that often plague people who are trying to lose weight.

Some people develop lightheadedness or a slight skin rash after taking chromium picolinate. If this happens, try cutting the dosage in half. If the condition persists, stop taking the supplement and see your doctor (see Chapter 10).

Given the fact that insulin function requires chromium, chromium is far more than a fat-reduction aid. You should continue taking chromium picolinate even after achieving your fat-loss goals.

7. CHROMIUM PICOLINATE AND DIABETES

There are two types of diabetes related to insulin production. In one type—the one associated with a lack of chromium—too much insulin is produced. In the other, little or no insulin is produced. A third type is not related to insulin production, and thus is beyond the scope of this book. (If you're not sure whether you have diabetes or not, see "How Healthy Is Your Insulin?" on page 65.)

In Chapter 5, we looked at how malfunctioning insulin leads to insulin impotence, and eventually, for some people, to diabetes. We also discussed the relationship between chromium and insulin. In this chapter, we will take a closer look at how chromium picolinate can help fight both kinds of insulin-related diabetes. We'll also discuss diabetes and pregnancy.

DIABETES CAUSED BY THE LACK OF INSULIN

Let's first consider the type of diabetes in which insulin can't be produced at all.

This type starts early in life, when, for some yet unknown reason, the body's immune system attacks and destroys the beta cells in the pancreas—the cells where insulin is made. Total destruction of these cells deprives the body of all insulin because no other organ produces this hormone.

It takes a period of weeks or months for total destruction to occur. As insulin production is reduced, the body's ability to store fuel in the form of fat is also reduced. This produces a number of symptoms. Soon after the raids on the beta cells start, the person will begin to look thinner. Very little insulin is being released from the pancreas, so more and more fat is detached from holding sites and released into the bloodstream. With little or no glucose and fat entering the body's cells, and with fat being used as fuel, the person wastes away. A second-century physician characterized diabetes as "being a melting down of the flesh and limbs into urine."

As the disease progresses, loss of fat and sugar continues. The diabetic soon starts urinating excessively and drinking gallons of water. Because there is little if any insulin to direct the glucose into muscle and fat cells, the amount of glucose in the bloodstream becomes very high. Glucose is important, so most of what is filtered in the kidneys normally is returned to the blood. In the diabetic state, however, the amount of glucose in the bloodstream is far too great for the kidneys to regulate. After breaking through the kidneys' filtering tubes, glucose pours into the urine, taking a lot of water with it to keep the glucose dissolved. Many vital nutrients also get carried along in the water and are lost in the urine.

The water accompanying the lost glucose causes the excessive urination. The loss of water from the body causes the thirst. When the excessive urination and thirst get bad enough, the affected individual will start to feel weaker and will eventually end up in a doctor's office.

Diabetes caused by beta-cell loss is sometimes called juvenile diabetes because it most often occurs before or during adolescence. It has also been called Type 1 diabetes. However, the name now preferred by doctors is insulin-dependent diabetes mellitus (IDDM). *Mellitus* means "honey-sweet" and refers to the high amount of glucose in the urine of people who have the disease.

Problems With Injected Insulin

Before insulin was discovered, a diabetic wasted away and died. Now this hormone is readily available, so a diabetic can start taking injections of insulin as soon as the disease is diagnosed. There is no such thing as oral insulin. Because insulin is a protein and proteins are destroyed in the intestines, the diabetic must inject insulin under the skin, and the injections must be continued for a lifetime. Insulin injections do not cure the disease. They merely provide a means for putting a vital hormone into the bloodstream.

Unfortunately, even though insulin keeps the patient from wasting away, injected insulin produces its own set of problems. It is not completely effective because it is not attached to chromium. As a result, blood sugar is not adequately controlled, causing the blood to have too much glucose and too much insulin at the same time.

Insulin turns on the mechanism for putting fat into fat cells, while high blood-sugar levels turn off the mechanism for releasing fat from fat cells (see Chapter 6). Soon after a diabetic starts injecting insulin, he or she often has problems with excess body fat. Too much fat is just one of many problems caused by the presence of too much insulin and too much glucose in the bloodstream. Both insulin and glucose are vital to life, but too much of either causes problems.

By Itself, Insulin Is Not Good Enough

Chromium plays an important role in alleviating some of the problems connected with too much insulin and glucose in the bloodstream. Whether or not insulin can act without chromium is not known with complete certainty. However, we do know with utmost certainty that insulin works more effectively in both humans and animals when they are given chromium supplements. In the Israeli study led by Dr. A. Ravina that we discussed in Chapter 2, IDDM patients showed an increased ability to remove glucose from their bloodstreams—an ability called glucose clearance—after only ten days of chromium picolinate supplements.[1]

Animal studies have proven that chromium picolinate supplements lead to normal or more rapid glucose clearance with much less insulin.[2–7] Many investigators have noted marked improvement in insulin action when chromium is used in test-tube experiments.[8–10] All of this evidence indicates that an IDDM patient can reduce the amount of insulin needed by taking chromium picolinate.

DIABETES CAUSED BY INSULIN IMPOTENCE

Although chromium supplements are beneficial for insulin-dependent diabetics, there is no evidence as yet to indicate that chromium can stop the destruction of insulin-producing cells and thus prevent IDDM. That is not the case with non-insulin-dependent diabetes mellitus (NIDDM), also called Type 2 or adult-onset diabetes. NIDDM is actually the final phase of what I have been calling insulin impotence. Chromium supplements,[11–18] particularly chromium picolinate, can do much to prevent the progression from insulin impotence to NIDDM.

NIDDM is a long-term disease that progresses through various phases. During the early phase, when a lack of chromium leads to a decrease in insulin efficiency, the main impairment is slow glucose clearance from the bloodstream.

As we saw in Chapter 5, chromium helps the insulin molecule keep its shape. If insulin is misshapen—a definite possibility when chromium is in short supply—it has a much more difficult time slipping through the cracks and crevices in the capillary walls. Misshapen chromium also has a difficult time attaching to the receptors on the surface of muscle and fat cells. If insulin can't find and attach to the receptors, then glucose can't enter the cells. As a result, the muscle and fat cells, which normally clear glucose after a meal, don't take up the newly digested glucose as quickly as they should.

If glucose can't enter the muscle and fat cells, it continues to circulate in the blood. Because the level of blood glucose stays high, the beta cells in the pancreas are tricked into believing that there isn't enough insulin in the bloodstream, and more insulin is released. This constant release of insulin from the pancreas goes on for a couple of hours after a meal.

Eventually, even misshapen insulin molecules can make it through the capillaries and attach themselves to the muscle and fat cells. But the presence of excess insulin means that almost every receptor has an insulin molecule attached to it, leading to a large rush of glucose into the cells.

The result is the clearance of far too much glucose from the bloodstream. The blood glucose drops from a very high level to a level that, while not life-threatening, is too low to completely satisfy the nerve cells in the brain. Since the brain needs a constant supply of glucose, the marked drop triggers an alarm in the satiety center, the part of the brain that monitors fuel supply. It sends out a low-fuel warning, causing the person to look for a quick infusion of sugar.

What Is Hypoglycemia?

Many people refer to what I just described as hypoglycemia, or low blood sugar. True hypoglycemia is a chronic condition in which the body has a difficult time keeping enough glucose in the bloodstream for normal brain function. It is not related to insulin impotence, and therefore is not covered in this book.

The condition we are concerned with is a temporary one, in which malfunctioning insulin causes blood-glucose levels to build up and then suddenly drop below normal. Doctors call this reactive hypoglycemia. A person who has this problem often goes from feeling anxious and hyperactive to feeling listless and depressed, and suffers from mood swings and food cravings. As a result, too much carbohydrate is eaten and the cells store too much glucose after a meal or snacking binge.

Dr. Richard A. Anderson and his colleagues at the Department of Agriculture have shown that chromium supplements are very effective in treating reactive hypoglycemia.[19] Chromium deficiency is difficult to detect at the cellular level but can be easily detected by changes in behavior. Drastic mood changes or abnormal cravings for any one type of food may be the body's way of calling for help.[20]

I have heard and read several reports from women claiming that severe symptoms of premenstrual syndrome were eliminated after they started taking chromium picolinate. Pharmacists have reported that chromium picolinate supplements decrease the amount of drugs needed to treat depression. A California doctor actually obtained a patent for the use of chromium picolinate in treating drug and alcohol addiction.[21] Since over 90 percent of the population is not getting enough chromium, there is a strong possibility that reactive hypoglycemia, and the chromium deficiency that helps cause it, might be the source of many behavior problems.

Too Much Glucose, Too Much Insulin:
The Development of NIDDM

The second phase of insulin impotence starts after the muscle and fat cells have been forced to gorge themselves with glucose for an extended period of time. The length of time the cells will tolerate the glucose onslaught varies, depending on heredity, diet, and lifestyle. Low sugar intake and exercise can delay the onset of the second phase for several years. On the other hand, high sugar intake and lack of exercise can hasten it.

As we saw in Chapter 5, muscle cells eventually ignore insulin's signal to take in glucose because glucose can gum up cell operations. This diverts glucose to the fat cells. In this phase, subtle but definite changes begin occurring through-out the body, especially in the blood vessels. Glycation—the attachment of glucose to various cell parts and chemicals—brought about by the higher glucose levels begins to take its toll, interfering with normal operations in the bloodstream and within the cells.

The second phase accelerates when the fat cells can no longer keep up with the glucose influx. They eventually become packed with stored fat and can't convert glucose into fat fast enough to prevent glucose accumulation. Like the muscles, the fat cells begin ignoring insulin's signal.

At this stage, there aren't many cells left to take up glucose. Even the liver, which can usually be depended on to remove various substances from the blood, is overcome. While the liver does have some glucose storage capacity, that capacity is limited. As a result, glucose floats around in the blood-stream for long periods after a meal. In this phase, insulin impotence has escalated into a disease characterized by hy-perglycemia, or high blood sugar. The high blood-glucose levels promote continued glycation in the blood vessels, es-pecially the tiny capillaries. This makes it more difficult for blood to flow freely through the circulatory system. Blood pressure goes up and it becomes more difficult to limit the amount of body fat.

Eventually, both the muscle and fat cells stop making glucose transporters, the protein tunnels that carry glucose from the surface of the cell to the interior (see Chapter 5).

Conservative structures that they are, cells don't bother making proteins that haven't been used for some time. This is the beginning of the next phase of insulin impotence. At this juncture, the pancreas continually gets a message to produce and secrete more insulin. The cells in the pancreas have no way of determining if the glucose sounding the alarm comes from a digested meal, or if it's just extraneous glucose looking for a place to go. Acting according to program, the beta cells keep pumping insulin into the bloodstream to stem the rising tide of glucose.

More insulin entering the bloodstream causes more insulin to combine with the insulin receptors on the membranes of cells. This leads to another problem—loss of insulin receptors. Since the cells are already clogged with glucose and fat, they have stopped producing glucose transporters. With no transporters, there is no reason for the cells to produce insulin receptors. Therefore, the cells stop manufacturing insulin receptors.

Prior to this point, the excess insulin attached to receptors on the cells. Even though the attachment was futile, insulin was taken out of the bloodstream. With fewer receptors on the cells, insulin is left to float around in the bloodstream. This leads to hyperinsulinemia, or high blood insulin. This extra insulin, or renegade insulin, has a number of unhealthy effects, including obesity, premature aging, and increased risk of heart disease. It more than likely contributes to a few other diseases.

In this phase, insulin impotence is a hyper, hyper disease—hyperglycemia coupled with hyperinsulinemia. That's a very unhealthy pair.

The epidemic disease caused by low dietary chromium progresses from misshapen insulin to a point where insulin can't combine with cells because they stop producing insulin receptors. At this stage the disease is referred to as overt diabetes mellitus, overt meaning that the symptoms of the disease are

A Word of Caution

Chromium picolinate is not a drug, but quite often it completely eliminates the need for drugs used to treat NIDDM, and can reduce the insulin needed by patients with IDDM. Patients with either NIDDM or IDDM who start taking chromium picolinate need to monitor their blood-sugar levels very, very carefully. Chromium picolinate increases the efficiency of injected insulin or drugs used in treatment. This could lead to hypoglycemic shock, in which blood-glucose levels drop too sharply, because more efficient insulin or drug action can cause too much glucose to be removed from the bloodstream. If you are a diabetic, I would strongly urge you to discuss the use of chromium picolinate with your doctor. Your doctor can help you find the right dosage of drugs or insulin to control your diabetes without triggering an adverse reaction.

very apparent. Blood-glucose levels are two or three times normal. At those levels, the kidneys run out of transporters to move glucose from the filtration tubes back into the bloodstream. As a result, the unreturned glucose stays in the filtration tubes until it gets dumped into the bladder. Glucose loves water, so it always takes a lot of water with it. Thus, sweet water is lost through the kidneys in copious quantities.

By this time—often sooner—the individual will become aware that something's wrong, and will have made a trip to the doctor. Once the doctor diagnoses the disease as NIDDM, various drugs will be prescribed to get the glucose and insulin out of the bloodstream. If the disease cannot be controlled with drugs, insulin injections are often prescribed even though the individual can still make insulin.

The progression of insulin impotence through the various stages can be prevented by chromium picolinate. Many studies with humans and animals have shown that chromium picolinate supplements make it possible to control blood sugar with less insulin production.[22–27] During a longevity study I conducted with my colleagues at Bemidji State University, rats given chromium picolinate lived longer and had more efficient insulin.[28] Controlling the removal of blood glucose with a minimal amount of insulin prevents the development of the myriad problems associated with hyperglycemia and hyperinsulinemia.

CHROMIUM PICOLINATE AND THE DIABETES OF PREGNANCY

During pregnancy, many women develop high blood-sugar levels. This, in turn, leads to high blood-insulin levels and excessive urination—a condition known as gestational diabetes. The studies of Dr. Merlin Lindemann in Virginia[29–31] (Chapter 3) and Dr. Lois Jovanovic-Peterson in California[32] (Chapter 2) show that chromium picolinate supplements can lower the high glucose and insulin levels that often occur during pregnancy.

The studies conducted by Dr. Lindemann and his associates have shown that gestational diabetes can be prevented with chromium picolinate supplements. In these studies, done with pigs, much less insulin was needed to clear glucose from the bloodstream when the test animals were fed chromium. The test animals were put on chromium picolinate when they were young, before they became pregnant. Therefore, I think that women should be encouraged to start chromium picolinate supplements as soon as they reach childbearing age, if not sooner.

Lindemann's studies also give evidence that the use of chromium picolinate can enhance the health of the child. Pigs fed chromium picolinate for over nine months—about three years in human age—had more and bigger offspring, and were more likely to have a second litter. Pigs fed even higher amounts of chromium picolinate, equivalent to nearly 1,000 micrograms of chromium a day in a human, until maturity also had healthy litters. I believe that these results also demonstrate chromium picolinate's safety as a supplement.

Dr. Jovanovic-Peterson's studies were done on pregnant women. In the first trial, the women received between 200 and 400 micrograms of chromium a day. That study showed good results, but another study, in which the quantity of chromium picolinate was doubled, showed even better results. Dr. Jovanovic-Peterson thinks that chromium picolinate can cure gestational diabetes.

USING CHROMIUM PICOLINATE
TO HELP TREAT DIABETES

Chromium picolinate cannot reverse the course of IDDM because it cannot stop the destruction of the beta cells in the pancreas. However, it can help the IDDM patient reduce the amount of insulin needed to control the disease. For people with NIDDM, chromium picolinate has proven to be an effective addition to insulin or drug treatment. Regardless of the type of treatment, chromium picolinate improves insulin efficiency.

No matter what type of diabetes you have, it is very important that you work with your doctor in finding your optimum dosage of chromium picolinate. Begin with 600 micrograms and move up towards 1,000 micrograms, stopping wherever you feel most comfortable.

Chromium picolinate can also be used to help prevent diabetes in pregnant women. If a woman has not been taking chromium supplements for at least a year prior to pregnancy, she may need 600 to 1,000 micrograms of chromium a day to prevent gestational diabetes. However, if supplements of 400 micrograms of chromium as picolinate have been used for at least a year, that amount should be continued.

Diabetics should read "A Word of Caution" on page 95 before using chromium picolinate. If you are a pregnant woman, mention chromium picolinate during your next pre-natal checkup, since it is always a good idea to keep your doctor informed of any change in habits. Some people develop lightheadedness or a slight skin rash after taking chromium picolinate. If this happens, try cutting the dosage in half. If the condition persists, stop taking the supplement and see your doctor (see Chapter 10).

Everybody needs chromium supplements because ordinary diets just don't contain enough bioavailable chromium. Individuals who have a family history of diabetes are prime candidates for chromium supplements because those people have a genetic makeup that somehow causes their insulin to become ineffective more quickly than usual.

8. THE ROLE OF CHROMIUM PICOLINATE IN SPORTS

W e've already seen how chromium picolinate can help the dieter lose fat. But this remarkable supplement can also help the athlete gain muscle, and do so without risking the serious side effects associated with steroid use.

Chromium picolinate's muscle-gain effects, like all its other effects, are related to chromium's effect on insulin (see Chapter 5). Chromium restores insulin's efficiency in moving glucose from the bloodstream into muscle cells, which means that less insulin needs to be secreted. With less insulin in the blood, production of DHEA, a hormone related to muscle growth, increases. Like all cells, muscles need a hormone stimulus before they will grow and develop. Therefore, chromium picolinate can help stimulate muscle growth. There are several studies that prove chromium picolinate's effectiveness in increasing the rate of muscle development.[1–8]

In this chapter, we'll look at the connections that tie chromium picolinate, insulin, and DHEA together, and why more studies are needed in this area. We'll also look at why steroid use is dangerous and no more effective than chromium picolinate use. Finally, there are some recommendations for using chromium picolinate to help build muscles naturally.

CHROMIUM PICOLINATE CAUSES
INCREASED STEROID PRODUCTION

I first read about a connection between insulin and dehydroepiandrosterone (DHEA) as part of my own research with chromium picolinate (see Chapter 1), and was especially interested to find that excess insulin shuts down DHEA production. DHEA is considered a weak androgen, which is a hormone used to create muscle proteins. DHEA can stimulate the development of muscle, but it's not as potent as testosterone, the most potent androgen. However, DHEA can readily be converted into testosterone.

I began to wonder if that particular hormone might help explain the observations, made by scientists and athletes alike, that the use of chromium supplements led to muscle development and fat loss. Could these results be linked to a drop in blood-insulin levels, a drop that led to better DHEA production?

To obtain some more information about DHEA production, I and my students at Bemidji State University (BSU) measured levels of blood insulin and DHEA in 12 men right after they ate a meal at a local fast-food establishment.[9] The men, all members of a health club, then exercised thirty minutes a day, three days a week. After four weeks, we again measured insulin and DHEA after a meal. The insulin level decreased by 11 percent and the DHEA level increased by 20 percent.

In this part of the study, exercise improved insulin efficiency even without chromium picolinate supplements. This isn't surprising, since exercise burns glucose, and too much glucose in the bloodstream can hinder insulin's effectiveness.

We then gave the men bottles of either chromium picolinate capsules or placebo capsules. (See "The Science of Studies" on page 7 for an explanation of how medical studies are conducted.) They were told to take two capsules a day and to continue exercising. After four weeks, we measured insulin and DHEA levels. We then switched capsules—the men who were receiving chromium picolinate now received the placebo, and vice versa. Again, they were told to take two capsules a day and to continue exercising for a final four weeks. Again, after-meal insulin and DHEA levels were measured at the end of the period.

At the end of the final four-week period, we discovered that chromium picolinate supplements have a pronounced effect on insulin and DHEA. As I said before, exercise alone caused insulin levels to decrease and DHEA levels to go up. However, during the four weeks the men took chromium picolinate, there was a big decrease in blood-insulin levels— 20 percent compared with 11 percent for the exercise-only phase. There was also a very large increase in DHEA—nearly 50 percent compared with 20 percent. On the other hand, insulin and DHEA levels during the four weeks the men took the placebo weren't much different from the exercise-only levels.

We also wanted to see if chromium picolinate use led to a faster rate of muscle growth. Lean body mass was measured at the beginning and again at the end of the study using a very accurate procedure known as biological impedance, which measures the ability of an electric current to pass through tissue. We found that muscle increased more when the men were exercising and taking chromium picolinate than when they were exercising without taking the supplement.

This study shows that DHEA levels go up and insulin levels go down when chromium supplements are used. The combination of chromium and exercise was especially beneficial in improving insulin efficiency. It led to less insulin in the bloodstream and a much greater output of DHEA. The elevation in DHEA, either by itself or by its conversion to testosterone, accelerated muscle development. The men gained 1.75 pounds of muscle with the combination of exercise and supplement but gained only a tenth of that with exercise alone. The studies provide strong evidence for a link between increased DHEA levels and improved muscle development.

We were also left with the sobering proof that supposedly fit athletes can still have some degree of insulin inefficiency. The studies show us there is indeed some room for improvement on Nature's design without using illegal and dangerous drugs.

LET YOUR BODY DO THE WORK NATURALLY

Bodybuilders and other athletes work tirelessly to develop muscle. All too often, progress is not made, so they turn to synthetic anabolic steroids. This is unfortunate because the liver turns anabolic steroids into estrogens, which leads to feminine characteristics in men and ovary shrinkage in women. The excess steroids can also cause kidney problems, liver cancer, heart disease, and abnormal behavior. The use of synthetic anabolic steroids to accelerate muscle development is totally unnecessary because the body is capable of making enough steroid to stimulate muscle development without such dangerous side effects. To accomplish this, the body's insulin has to work efficiently.

Muscle development requires four basic materials:

- Glucose
- Fat
- Amino acids
- An activation hormone

Glucose and fat are used as fuel to produce energy within the muscle cells. Amino acids are needed to make the proteins that form the muscles. And an anabolic, or growth-promoting, hormone is needed to switch on the process that turns amino acids into proteins.

A lot of research points to the role of insulin in determining whether or not all of those materials reach muscle cells when they're needed. Malfunctioning insulin slows muscle development and promotes fat formation. When insulin doesn't work, muscle cells don't get enough fat, their preferred fuel, because the fat can't get out of the fat cells (see Chapter 6). When muscle cells can't use fat, they turn to glucose for the fuel they need to produce energy.

However, as we saw in Chapter 5, cells do not take in glucose when insulin malfunctions. If that happens, glucose can't be used for energy. If fat can't be readily released from fat cells and glucose can't get into muscle cells quickly and efficiently, the muscle cells are then forced to convert amino acids, the protein building blocks, into energy.

Malfunctioning insulin keeps these vital building materials from getting inside the cells. Those that do are thrown into the cells' furnaces to generate energy. Sometimes, the muscles will tear apart their own proteins to obtain fuel. This is what happens when an athlete can't seem to go beyond a certain point in muscle development. Chromium keeps insulin working at maximum capacity so that muscle cells can use glucose and fat for energy, and use amino acids to make protein.

Proper use of glucose, fat, and amino acids certainly contributes to accelerated muscle development, but there is also another very important factor involved—the activation hormone, or androgen, needed to stimulate the manufacture of muscle protein.

Muscles don't just make protein on a whim. They require a hormone signal. When muscles are exercised regularly, the

Chromium Picolinate Brings Out the Polygraph

Deborah Hasten, Ph.D., studied the effect of chromium picolinate on athletes as a graduate student at Louisiana State University. She not only has a professional interest in the subject, but a personal one as well. Dr. Hasten is a bodybuilder who uses chromium picolinate herself, and it was through the sport of competitive bodybuilding that she realized the true power of this important supplement.

In 1992, Dr. Hasten entered a contest in Louisiana, where she was the obvious winner because of her superior, well-defined physique. However, the judges suspected her of using synthetic anabolic steroids, which are banned in such competitions. They didn't believe her when she instead credited her build to chromium picolinate, which at the time, was not as well-known on the bodybuilding circuit as it is today. The judges were so suspicious that they wouldn't give her the first-place prize until she took a polygraph test. She passed, of course, and swept the competition.

brain causes testosterone, the most potent androgen, to be secreted into the bloodstream. Testosterone stimulates muscle development, but the amount of testosterone produced by the body is limited.

To stimulate muscle development beyond the limits of testosterone production, a backup is needed: DHEA. DHEA can stimulate the development of muscle. It isn't as potent as testosterone, but it is an androgen. Also, DHEA can readily be converted into testosterone. And as we've seen, the body produces more DHEA when there is less insulin in the bloodstream, and the amount of insulin in the bloodstream is determined in large part by the amount of chromium available to help insulin function effectively.

Therefore, taking anabolic steroids is not only unsafe, it is unnecessary. What's needed is chromium to ensure that insulin functions properly. When insulin is working efficiently, so that there isn't excess insulin slowing down DHEA production, the body is capable of producing enough anabolic hormone to stimulate muscle development.

SIMILAR CHEMICALS
HAVE MUCH DIFFERENT ACTIONS

After we announced the results of the studies described in Chapter 1—the studies that showed how chromium picolinate use, insulin efficiency, and muscle development are related to each other—many scientists and people in the health-food industry wondered if chromium picolinate is truly different from other forms of chromium. Most, but not all, accepted the fact that chromium salts aren't effective. But they wanted to see some comparisons with forms of chromium that are more like chromium picolinate.

In a study with humans at BSU, we compared the effects of chromium picolinate and another form of chromium called chromium nicotinate on body composition, or the balance between muscle and fat.[10] For these tests, 12 men and 12 women, aged twenty-five to thirty-six, in a weekly aerobics class were divided into two groups of 6 men and 6 women each. Each of the men was given a bottle of capsules, each containing 400 micrograms of chromium as either chromium picolinate or chromium nicotinate. The women were given 200 micrograms of chromium. The participants were instructed to take one capsule each day at breakfast. Lean body mass was determined using biological impedance at the beginning of the exercise regimen and again after twelve weeks.

At the end of the study, muscle mass had increased in everyone, but the increase was three times greater in both men and women who took chromium picolinate.

In a study headed by Dr. Robert Lefavi at Georgia Southern University, male weightlifters were given supplements of chromium nicotinate.[11] The supplement resulted in lower blood-cholesterol levels but did not produce any significant changes in body composition. Obviously, chromium nicotinate and chromium picolinate are used in different ways by the human body.

At BSU, we also compared the effectiveness of six different forms of chromium on muscle cells grown in test tubes.[12] As long as insulin was in the test fluid, all the cells took in glucose. The muscle cells grown in fluid containing chromium tripicolinate, the form of chromium picolinate you buy in the store, took in about twice as much glucose as the cells grown with the other forms of chromium. When no insulin was put in the test fluid, the cells did not take in glucose. This proves that none of the chemical forms of chromium, including chromium tripicolinate, had any effect without insulin, but that chromium tripicolinate was much more effective than the other forms of chromium in helping insulin.

We also used zinc picolinate and picolinate by itself. The results showed that neither substance had any effect on the amount of glucose taken in by the muscle cells. This proves that the effect on insulin is not caused by picolinate itself, but by chromium.

WANTED: MORE STUDIES

Since 1989, chromium picolinate has become a well-known and fairly popular supplement among athletes. This fact makes it extremely important that new studies be done with people who have not used chromium picolinate for at least three months.

For example, in one study with male athletes,[13] the investigators found no difference in muscle development between men given chromium picolinate and men given a placebo. During that study, the researchers collected urine and determined the chromium content at the beginning, the middle, and the end of the study. At the beginning, the average chromium content of the urine from men who had been given a placebo was nearly twice the amount of chromium in urine from the men given chromium picolinate. By the end, the men given the chromium had greatly increased amounts of chromium in their urine, as expected. But the amount of chromium in the urine from men given the placebo dropped by half. Because this was a double-blind study, in which neither the participants nor the researchers knew who was taking what until the end, the relationships between chromium levels and substances taken were not clear until the study was over.

What could explain these results, especially with a substance like chromium that is not present in the diet? Is it possible most of the men who were given the placebo had been taking supplements that included chromium? If so, they obviously stopped during the study because their urinary chromium levels went down. But if they had been taking supplements, possibly even chromium picolinate, how much effect would that have had on their body chemistry? Were the men on the placebo in fact benefitting from their previous supplements, enabling their muscle development to keep pace with that of the men on the chromium supplement? These are the kind of problems investigators face when doing studies with human beings.

Studies with humans run into other problems. For example, in one study, 16 men were enrolled in a training program. They were divided into two groups of 8 each. One group took 200 micrograms of chromium a day in picolinate form, and the other took a placebo. After twelve weeks, the lean body mass of the men using chromium increased by 2.2 pounds, while muscle mass of the men using the placebo increased by only 0.44 pounds. The men taking chromium lost fat equivalent to 1.0 percent of body weight, but the men taking the placebo gained fat equivalent to 0.8 percent of body weight.

Obviously, chromium picolinate was far more effective than exercise alone. The study's authors, however, stated that they found no statistical increase in muscle mass in either group, and dismissed chromium picolinate as ineffective.[14]

I believe that this study's design was flawed because there were not enough participants to provide meaningful results, and because the dosage of chromium picolinate was too low. I also believe that the results were not properly interpreted because overly strict statistical standards were used.

Because of chromium picolinate's potential for increasing muscle mass and improving athletic performance, far more studies with this nutrient are needed. Animal studies have provided some guidelines that need to be followed in designing human trials. In the animal studies, chromium was added to the diets so that chromium intake was in step with calorie intake. Food intake was measured throughout the studies.

Most human studies—including, I admit, our first studies—are designed with little or no regard for calorie intake. calorie intake is critical because it reflects physical activity and muscle mass. More activity means greater mass. Based on the observations in animals, no human study using less than 400 micrograms of chromium a day should even be considered relevant, especially when athletes are involved.

Studies with athletes, although necessary, are ripe with unknown factors—prior and present diet, diseases, family history, activity level, locale, mood, time of month and year. Designing a human study that takes all these factors into account is nearly impossible. More studies, designed by scientists who specialize in the effects of exercise on the human body, are needed.

USING CHROMIUM PICOLINATE
TO HELP GAIN MUSCLE

While there is a need for more research, I believe that we already have strong evidence of a link between chromium picolinate and muscle gain. Therefore, most athletes will find that between 600 and 800 micrograms of chromium in picolinate form is very helpful for increasing muscle mass and losing fat. It is also likely that performance will also be enhanced.

Chromium picolinate supplements help develop muscle, especially in people who exercise, so it seems logical that athletic performance should improve. Unfortunately, nobody has done any tests to measure changes in performance because the issue is not as simple as it seems. Many factors must be taken into account when designing such a study.

I have gotten many letters from athletes who claim that they had noticed a big difference in their performance after taking chromium picolinate, and wanted to know how it works. I always encourage them to start keeping a record of their performance so they can see if the supplement makes any difference. You should do the same, especially if you are already fairly competent in your sport and are just beginning to use the supplement. If you are just starting in some athletic endeavor, you will improve simply because you will become more experienced. The best measure of performance is in athletes who think they have reached a peak and can't progress any further.

Start by assessing your performance now, and then keep a record of how you do while you are taking chromium picolinate. Keep in mind that changes in performance may not always be easy to measure. A weightlifter may actually find that he or she can lift more weight. On the other hand, that person might realize that although no increase in weight can be achieved, he or she can tolerate much longer workouts in the gym. Whether you are a tennis player, a marathon runner, a cyclist, or a swimmer, watch for changes that are subjective—how you feel—as well as objective—what you can measure.

Some people develop lightheadedness or a slight skin rash after taking chromium picolinate. If this happens, try cutting the dosage in half. If the condition persists, stop taking the supplement and see your doctor (see Chapter 10).

I hope that some unbiased athletic trainer will someday design a valid study—which will be difficult—and provide us with some scientific evidence of chromium picolinate's effect on athletic performance, either for or against. In the meantime, remember that synthetic anabolic steroids are both dangerous and unnecessary. Try chromium picolinate and measure your performance, and make your own judgement about changes objective or subjective.

9. USING CHROMIUM PICOLINATE TO SLOW THE AGING PROCESS

S tudies show that chromium picolinate prevents fat accumulation, promotes muscle development and fat loss, lowers blood-cholesterol levels, controls blood-sugar levels, and helps prevent osteoporosis, just to name a few benefits. All of the positive effects of chromium picolinate can be traced to its effect on insulin efficiency (see Chapter 5). Since improving insulin efficiency prevents many potentially serious health problems, it shouldn't come as a surprise to learn that there is a link between insulin efficiency and preserved youth.

In this chapter, we'll look at how inefficient insulin speeds up the aging process, and how chromium picolinate can slow it down. We'll also look at chromium's effectiveness in combatting both high blood-cholesterol levels and osteoporosis, two conditions strongly associated with aging. Finally, we'll discuss some recommendations for using chromium picolinate as part of an anti-aging program.

CHROMIUM IS THE LIFE SUPPORT SYSTEM WE NEED

A longer life in itself is not really what most people desire.

Nobody wants to spend their last years in a nursing home connected to a life-support system just so they can stay alive. Advances in medicine have made it possible for us to live longer—the life span of humans in the industrialized world has increased to about seventy-five years. But now we need a way to stay healthy longer. People want to preserve their youth.

In this section, we'll see how inefficient insulin quickens the aging process throughout the body, especially in the master glands that control all of the body's major functions. We'll also see how chromium picolinate, through its effects on insulin, can help the body stay young.

Insulin, Chromium, and Aging

Why does the body age? One view is that glycation—the process by which glucose attaches to various cells and chemicals—is a primary cause of aging.[1] Glycation of proteins and other substances is most common in diabetics, but it occurs whenever glucose levels become higher than Nature intended. Insulin malfunction leads to elevated blood-glucose levels, which eventually causes glycation to occur.

To test this idea, a research group at Texas A and M University fed rats a calorie-restricted diet, a diet that contained 60 percent of the calories rats normally eat.[2] This approach was used because it had been discovered during the 1930s that rats fed calorie-restricted diets lived much longer.[3-4]

The Texas group found that the rats not only lived longer, but also had lower blood-glucose levels. What's more, the rats had much less glycated hemoglobin in their bloodstreams. Glycated hemoglobin is a combination of glucose and hemoglobin, the chemical that carries oxygen in the blood. It is an indicator of excess glucose in the bloodstream.

In one of the Mercy Hospital studies described in Chapter

1, I and my colleagues had seen lower levels of glucose and glycated hemoglobin in the blood of diabetics who had been given chromium picolinate.[5] Obviously, there was something about both calorie restriction and chromium supplements that led to lower levels of glucose and glycated hemoglobin, which in turn led to longer life. What was it?

To answer that question, I and my students at Bemidji State University (BSU) decided to feed chromium to rats and see if that would increase the rats' life spans.[6] In this study, described in Chapter 3, we found that rats fed chromium picolinate had markedly reduced levels of glucose and glycated hemoglobin in their bloodstreams. But we also found that the rats fed chromium picolinate had lower insulin levels, and that less insulin led to longer lives.

There are undoubtedly many factors involved in the aging process, but chromium and insulin are definitely among them. The hormone DHEA, which we've mentioned in Chapters 1 and 8, has been shown to have several anti-aging properties,[7] and we know that DHEA production is decreased when there's a lot of insulin in the bloodstream.[8] Protein glycation, which probably speeds the aging of cells, is related to insulin efficiency, because if insulin is functioning properly there is less glucose bumping around the bloodstream, and therefore less glucose that can attach to protein molecules.

Because insulin malfunction—what we've called insulin impotence—alters many body functions, a combination of insulin-related factors probably leads to cell degeneration. Insulin is not completely efficient unless there is enough chromium in the body. Chromium helps cells respond to insulin quickly and effectively. This mineral obviously plays a key role in warding off disease, and thus helps keep the body healthy and vigorous.

To preserve youth, one must first preserve cells. Cells, the basic building blocks of the body, work together to form tissue. Tissues work together to form an organ, and organs work together to form an organ system. Organ systems working in unison make up a living body.

Aging occurs because cells stop working at maximum capacity, a process that starts at about age thirty. When one cell stops performing, the reverberations are felt, over a period of years, throughout the body. The heart becomes stiffer and less efficient. Blood vessels become less flexible and resist blood flow, which increases the heart's workload. The less efficient heart and the decreased blood flow lead to fewer nutrients and more waste matter throughout the body. The lungs receive less blood, so the ability to take in oxygen and push out carbon dioxide diminishes. Viruses, bacteria, and tumors spread and reproduce because the immune system can't keep up with the more virulent intruders.

We notice these effects more and more as time goes on. Our eyesight and hearing aren't as acute as they were when we were younger. Our skin wrinkles, our hair turns gray. We move more slowly and stiffly. We find that our short-term memory isn't what it used to be.

Aging is never restricted to one site in the body. Once it starts, it progresses until one of the organ systems becomes too debilitated to endure any longer and simply succumbs, taking the rest of the body with it.

Aging: The Control Center

Aging is a complex process, although as we grow older it seems to happen a little too easily. We don't know all the factors that make our cells slow down. But many studies point to changes occurring in three critical areas:

- The hypothalamus
- The pineal
- The thymus

These glands secrete chemicals that are vital for cell maintenance. Their function, like so many other bodily processes, is affected by the efficiency of insulin, and so they too depend on an adequate supply of chromium in the body.

The Hypothalamus

We first mentioned the hypothalamus and its role in fat loss in Chapter 6. It is a collection of cells in the brain designed to function as both nerve cells and as an endocrine, or hormone-secreting, gland at the same time. Nerve cells can only communicate with adjacent nerve cells, but endocrine glands can communicate with various sites in the body by sending chemical messages through the bloodstream.

The hypothalamus is the link between the brain and the environment, both inside and outside the body. Any change occurring in the environment is detected by nerve cells that transmit the nature of the change to the hypothalamus. The hypothalamus then evaluates the situation and, when necessary, alerts the appropriate areas of the body.

In addition to regulating the body's hormones, the hypothalamus controls the involuntary, or autonomic, nerves. These nerves regulate a wide range of functions, such as temperature, hunger, and blood pressure.

Since the hypothalamus is the control center for the body, its aging has widespread effects.[9–12] In order for the hypothalamus to act as an early warning system, it must effectively sense the slightest change in the environment. An aging hypothalamus can't carry out those duties efficiently, just as an old thermostat can't detect a change in the air temperature nearly as efficiently as a new thermostat can.

All glands need stimulation by either hormones or nerve-transmitted chemicals. If stimulation decreases, or ceases completely, the gland itself—as well as the cells that are targeted by the gland's secretions—will slow down and shrink, and may eventually stop functioning altogether. The production of certain hormones and nerve-transmitted chemicals by the hypothalamus decreases with age. There is a corresponding decrease in the activity of several hormone-producing glands throughout the body, including the cells in the pancreas that secrete insulin.

The Pineal

The hypothalamus may be the master gland, but there is a gland that controls the hypothalamus: the pineal. This gland, located at the base of the brain near the hypothalamus, produces melatonin. Melatonin lets the hypothalamus know when it must readjust its secretions to compensate for activity affected by light, such as day/night cycles and seasonal changes.[13] Melatonin secretion slows with age—it has been found that injections of this hormone or implants of young pineals into old animals prolongs life. But the hypothalamus also maintains the pineal. When age slows nerve transmissions from the hypothalamus to the pineal, the pineal begins to shrink.

The Thymus

The third gland linked with the aging process is the thymus.[14] The thymus, located under the breastbone, enlarges during childhood as it helps the lymph nodes to start producing certain infection- and cancer-fighting white blood cells called T cells, and then starts shrinking soon after puberty. It is where the T cells learn how to defend the body against attack.

The fact that diseases of the hypothalamus cause accelerated shrinkage of the thymus indicates a link between the two. The pituitary, another master gland directly regulated by the hypothalamus, also affects the thymus. The thymus, in turn, protects the pituitary.

Chromium Keeps the Age Trio Young

A great deal of scientific evidence indicates that the hypothalamus, the pineal, and the thymus are interdependent. Deterioration in one of these vital glands promotes aging or loss of youth-retaining characteristics in the other two. Each of these glands is affected by the action of insulin. Each is therefore subject to the damage inflicted by insulin impotence caused by lack of chromium.

This interdependence is shown in study after study. The production of several hypothalamus and pituitary hormones is decreased in rats made diabetic by injection of a substance that destroys the beta cells.[15] The lack of insulin caused by diabetes also causes an accelerated degeneration of nerve cells in the hypothalamus.[16] Hormone production in both the thymus and the pineal is decreased in diabetic rats.[17] Almost all human diabetics generally have poorly functioning immune systems, which is probably related to a prematurely aging thymus.[18]

Whether or not lack of insulin or impaired insulin function directly alters the activity of one or more of these glands is not yet known. The fact is, when one of the age-affecting glands deteriorates, the trio goes together, and youth slips away.

Several years ago, Dr. Henry Schroeder, one of the first scientists to understand chromium's importance, proved that chromium deprivation shortened life spans in rats and mice.[19] More recently, I and my students at BSU extended the life spans of rats by adding chromium picolinate to their diets.[20] These longer-lived rats had less glycated hemoglobin, less glucose, and less insulin in their bloodstreams. These experiments prove that chromium—in a form that's readily absorbed by the body—prevents glycation and increases insulin efficiency. And the body needs to do both jobs if it is to remain youthful.

CHROMIUM PICOLINATE AND HIGH CHOLESTEROL

One of the things that happens as most people age is that their blood-cholesterol level rises, which can lead to heart disease. The connection between chromium and cholesterol was first discovered in the 1960s,[21-23] and has since been documented in several studies.

What is the connection between cholesterol and heart disease? You have probably heard about "good" and "bad" cholesterol, which makes it sound like there are two kinds of cholesterol. Actually, there is only one cholesterol, but it combines with different kinds of proteins. Cholesterol is a fat that cannot be dissolved in water. Thus, it has to be carried by proteins from the intestines to the liver, the organ that controls cholesterol use, and from there to other places in the body. The proteins that carry cholesterol throughout the body are called lipoproteins (*lipo* means fat).

The kind of protein the cholesterol is attached to determines whether it is good or bad cholesterol. Low-density lipoprotein (LDL) is the bad cholesterol that builds up on the walls of blood vessels. This buildup reduces blood flow. If a blood clot gets caught in the clogged vessel, a heart attack or stroke can result. High-density lipoprotein (HDL) is the good cholesterol that can clean up the bloodstream, resulting in a reduced risk of heart disease.[24-26]

A number of studies have shown that chromium picolinate can reduce levels of both total cholesterol and LDL cholesterol in the blood. In one study with human volunteers, total cholesterol decreased by 7 percent and LDL cholesterol decreased by 10.5 percent.[27,28] In another study, which used both chromium picolinate and the vitamin niacin, total cholesterol dropped by 24 percent and LDL cholesterol dropped by 27 percent.[29] And in another study, this one done with chickens, LDL cholesterol decreased while HDL cholesterol increased.[30]

Since there is a known link between cholesterol and heart disease, and a known link between chromium and cholesterol, it would be logical to assume that there's a link between chromium and heart disease. This has been confirmed by research.

In the mid-1970s, Dr. Howard Newman and his colleagues at Ohio State University studied chromium and heart disease.[31] Dr. Newman had read Dr. Schroeder's accounts of a possible connection between chromium and heart disease (see Chapter 1), and decided to launch his own study.

The Ohio State group analyzed blood from patients who had coronary artery disease, as well as blood from patients with other kinds of health problems. They found the lowest chromium levels in the patients with coronary artery disease. The study was particularly convincing because the analyses were painstakingly scrupulous and totally without bias. These tests showed that people who have a blood-chromium concentration at or below a certain level have a high probability of contracting heart disease. The findings of Dr. Newman's team were verified by a group of French scientists, who also found that the risk of heart disease is greatest when blood chromium falls below a critical level.[32]

Why does chromium lower cholesterol levels, and thus help prevent heart disease? Chromium controls insulin. As we saw in Chapter 6, insulin controls the body's use of fat, as well as its use of sugar. High blood-insulin levels results in a lot of fat being released from the proteins that carry fat through the bloodstream. A lot of that fat enters the fat cells, but a good share is also taken up by the liver.

This starts an unhealthy cycle. The liver is very efficient at dismantling unwanted chemicals. All that fat coming into the liver is not necessary, so the liver produces enzymes that break the fat into small pieces. Those small pieces of fat can be used to make cholesterol, which the liver does, in increasing amounts.

The liver must get the cholesterol into the bloodstream, and does so by packaging it with fatty acids and a certain kind of protein. This creates the LDL cholesterol we discussed earlier. This type of cholesterol is very sensitive to attack by chemicals called oxidants in the bloodstream. The oxidants change the protein and it ends up, along with its cholesterol and fat load, inside cells that line the blood vessels. Eventually, these cells become bloated with cholesterol and end up blocking the flow of blood.

Thus, chromium helps decrease cholesterol levels by reducing the amount of insulin, which in turn reduces the amount of fat in the bloodstream, which in turn reduces the amount of fat that is used to create cholesterol.

The Ohio State and French studies give us a means of predicting the risk of heart disease in people. Unfortunately, blood chromium analyses are still difficult and expensive. But when they become easier and cheaper, they can be used to determine the chromium needs of people who run a high risk of developing heart disease.

CHROMIUM PICOLINATE AND OSTEOPOROSIS

Older women often suffer from osteoporosis, in which their bones become brittle because calcium is lost through the urine. This is accompanied by a loss of height and what has come to be known as a dowager's hump—stooped shoulders caused by the backbone's inability to support the weight of the upper body. As we first saw in Chapter 1, osteoporosis occurs in postmenopausal women mainly because of a decrease in production of female hormones. Without enough of the main female hormone, estrogen, calcium is filtered through the kidneys.

Bone, like all other bodily tissues, is constantly being renewed. During childhood and adolescence, bone manufacture is far ahead of bone breakdown. As people reach their twenties and thirties, breakdown and rebuilding are equal, but by middle age loss is ahead of gain. In old age, loss of bone is far ahead of the rebuilding process, especially in women. Thin, porous bone—osteoporosis—affects one of every two women and one of every eight men.

We know that the hormones estrogen and testosterone help maintain bone by slowing down the cells responsible for breaking down bone tissues. Both of these hormones decrease with advancing age. Unfortunately for women, the loss of estrogen at menopause is much more rapid and dramatic than the loss of testosterone in men. That explains the higher incidence of osteoporosis in women.

Osteoporosis may also be linked to DHEA, a hormone that has shown other anti-aging properties. In women, DHEA can be converted into estrogen. However, there isn't enough DHEA in the blood of older women because most have inefficient insulin. Malfunctioning insulin, or insulin impotence, leads to higher blood levels of insulin. Insulin turns off the DHEA manufacturing process. Therefore, most menopausal women not only lose their ability to make estrogen, they also lose the ability to make the DHEA that could make up for the shortage. Studies show postmenopausal women have higher blood-insulin levels and much lower DHEA levels than women twenty years younger.[33]

Insulin affects bone density in other ways besides its effect on DHEA. Insulin receptors—the protein docks to which insulin molecules attach themselves—have been found on the cells where new bone is manufactured. Decreased bone mass has been observed in diabetics.[34] And properly functioning insulin also slows aging of the hypothalamus, the part of the brain that controls hormone production, including the production of estrogen.

I and my students at BSU conducted a study on insulin, DHEA, and estrogen (see Chapter 1).[35] We found that women who took a daily chromium picolinate supplement had

DHEA and estrogen levels normally found in women in their early thirties. The participants also retained much more calcium.

Ordinarily, osteoporosis is treated with estrogen or a combination of estrogen and calcium supplements. Estrogen is not an ideal treatment because it has been shown to increase the chance of certain kinds of cancer. A more desirable method of treatment would be a combination of chromium picolinate and calcium supplements.

USING CHROMIUM PICOLINATE TO HELP SLOW THE AGING PROCESS

Many factors cause aging of the body's cells, but proper diet and exercise, along with vitamins and minerals, are very effective in slowing the process. You should obtain about a third or less of your daily calories from fat. Also, foods that are rich in sucrose, a common sweetener, should be avoided because sucrose causes chromium to be washed out of the body (see Chapter 5), which leads to insulin impotence and protein glycation. Alcohol should be consumed in moderation, and smoking should be completely eliminated because tobacco smoke contains many chemicals that damage cells.

There are many benefits to be gained from exercise. A person who doesn't exercise doesn't use many muscles. Since a lot of amino acids and carbohydrates are needed to maintain muscle cells, the body tends to replace muscle cells with fat cells, which need hardly any of these substances. Muscle is valuable because it burns the fuels, glucose and fat, that are otherwise deposited in various areas of the body, such as the abdomen and thighs.

A daily vitamin supplement will help slow aging, but there are some vitamins and minerals that are particularly beneficial:

- 400 micrograms of chromium as picolinate

- 5 milligrams of zinc as picolinate
- 5,000 international units of vitamin A
- 500 milligrams of vitamin C
- 400 international units of vitamin E
- 10 milligrams of beta-carotene
- 20 milligrams of bioflavonoids
- 200 micrograms of selenium from yeast or as se-lenomethionine
- 1,200 milligrams of calcium

You may not recognize some of the specific mineral forms recommended. Like chromium and zinc, which are best utilized when attached to picolinate, the antioxidant selenium is most effective when derived from yeast or when attached to methionine. Also, vitamin C works much more effectively when bioflavonoids are present, and beta-caratone is another potent antioxidant.

Some people develop lightheadedness or a slight skin rash after taking chromium picolinate. If this happens, try cutting the dosage in half. If the condition persists, stop taking the supplement and see your doctor (see Chapter 10).

Obviously, no one lives forever. But why not live the healthiest life possible? An anti-aging program that includes chromium picolinate can help you enjoy a hearty old age.

10. CHROMIUM
PICOLINATE FOR LIFE

Insulin, as you now know, does more than just control blood sugar. Because insulin affects so many bodily activities, insulin malfunction can cause a number of diseases. But chromium, because of its effect on insulin, can prevent many of these diseases and conditions, such as non-insulin-dependent diabetes, heart disease, osteoporosis, obesity, muscle atrophy, and premature aging. Chromium is good news.

In this chapter, we will take a final look at the various ways in which chromium can help the body, and we'll review why that help must come from supplements. We'll also discuss the safety of chromium picolinate. Finally, I'll make some general dosage recommendations.

CHROMIUM THE FAT FIGHTER

I believe that chromium picolinate's primary role is to make insulin effective. Effective insulin, in turn, will serve as a fat-reduction aid.

Obesity—defined as body fat exceeding 30 percent of total weight in women and 20 percent in men—is a major health problem. A few years ago, 25 percent of the United States population was categorized as overweight or obese. That number has now increased to 33 percent. Obesity is considered a risk factor for heart disease, diabetes, hypertension, and possibly cancer.

Fat-control problems keep increasing despite widespread attempts to control body fat. Unfortunately, the treatment of obesity is most often ineffective. Of those starting a weight-loss program, 95 percent go back to their original weight within five years.

So what keeps people from staying lean all of their lives and fighting fat most of their lives? It probably has a lot to do with the fact that Americans eat about 133 pounds of sugar a year, but get only a small fraction of the chromium they need to combat that excessive sugar intake.[1] Sugar promotes the development of insulin impotence and accelerates the loss of insulin's helper, chromium (see Chapter 5).

I am a nutritional biochemist. As a university teacher, I often have my students do computerized dietary analyses, in which the students record what and how much they eat over a five-day period. The computer then prints out what nutrients each student should be consuming versus what he or she actually is consuming. The students also describe their activity levels so that the computer can compare calorie intake with output.

In the more than 500 analyses I have discussed with students, calorie intake is most often equal to, and very often less than, calorie use. So far, so good. But then the students do body-composition estimates—estimating the amount of fat on their bodies versus the amount of muscle. Imagine their shock when many of them learn that they are considered obese or nearly obese, despite the fact that their energy intake is just about balanced by their energy output.

Why Do People Become Obese?

I use several different textbooks when preparing my lectures. In nearly all of them, the author discusses how John and Jane Doe can lose weight. This fictitious couple usually consumes about 200 calories a day more than they use. The authors then go on to show how, by carefully counting calories, John and Jane can reduce their intake and thereby lose a couple of pounds a week. The authors fail to point out that reduced calorie intake will very often cause these people to feel weaker and less energetic, and very often, that they will gain back more fat than they lose.

The authors of these books go into great detail about many things, but they never address the real question: why did John and Jane become obese in the first place? Why do many of my students have too much fat on their bodies? Why do many other people in this country have the same problem?

The easiest way to explain the students' dilemma is to say that dietary analyses are not totally accurate and that the students are actually consuming more calories than they are using. That might account for the fat accumulation, but, as we saw in Chapter 6, high blood-glucose levels are supposed to activate the brain's satiety center so that people won't take in excess calories. And what about John and Jane? Most authors describe the satiety center but fail to explain why it doesn't work in their case.

One could argue that the satiety center is set a little higher in some people, so that it takes them longer to feel full. That is a definite possibility. But when more food than needed is eaten, thermogenesis—the body's ability to produce heat—is supposed to burn up the excess energy.

There is also the definite possibility that the basal metabolic rate (BMR)—the energy consumed by the body's basic functions, such as breathing—is lower in some people, so they don't use as much energy to keep the body operating.

However, if a lower BMR is causing less fuel to be used, why aren't the satiety center and thermogenesis preventing a buildup of fat?

A lack of chromium is certainly not the only cause of obesity or problems with body fat. However, considering how many people lack chromium in their diets, there is a strong possibility that inadequate dietary chromium contributes to the increasing incidence of obesity in this country. I could fill hundreds of pages with letters from people describing how chromium picolinate has helped them lose fat.

I am convinced that controlling insulin impotence is at least one answer to the dilemma of controlling body composition. Insulin impotence keeps glucose and amino acids from moving efficiently into muscle cells, and it keeps fat from being used as a primary fuel for muscle. When insulin malfunctions, the flow of fuel is away from muscle and toward fat cells. And insulin malfunctions when the body can't get enough chromium.

Overrated Exercise and Calorie-Counting Folly

Chromium's effectiveness in controlling body fat has been shown in many studies. Pigs, rats, lambs, and turkeys all accumulate less body fat when chromium is added to their daily rations. In one human study led by Dr. Gil Kaats, nonexercising volunteers lost over four pounds of fat without markedly changing their diets.[2] They simply took a chromium supplement every day. The fat loss had to result from increased thermogenesis as well as a higher BMR. Another Kaats study confirmed this assumption.[3]

Muscular activity—exercise—uses the food energy that isn't used by basic bodily functions. One study reveals some very enlightening facts about the effects of exercise on fat loss.[4] The authors list the results from forty-eight studies on men, in which volunteers who exercised for thirty minutes at least three times a week lost an average of 3.6 pounds. The average length of the studies was slightly over sixteen weeks. The authors also list the results of sixteen studies on women who exercised about twenty-five minutes a session four times a week for twelve weeks. Their average fat loss was 2.2 pounds. After weeks of strenuous aerobic exercise, the men and women participating in these studies hadn't lost as much fat as the nonexercising men and women in Dr. Kaats' studies.[5]

Exercise is healthy and it can lead to the loss of body fat, but only if the whole body is healthy. Chromium is definitely an effective fat fighter.

Decreasing calorie intake to balance output is foolish and unhealthy. Even if you are consuming too many calories, you still aren't getting enough of the important vitamins and minerals you need. Cutting down calorie intake will only make the problem worse, since you will feel weak and hungry. Remember, if your cells are insulin resistant, it's very difficult to move fat from its storage sites into the muscles.

Not everyone is going to get rid of excess fat merely by preventing insulin impotence. That would indicate obesity can be eradicated simply by having enough chromium in the diet. However, if you want to lose fat or prevent fat from accumulating, you should take steps to improve your insulin function by getting adequate chromium into your system.

CHROMIUM: THE ALL-STAR NUTRIENT

Given the number of people in this country who struggle with weight problems, chromium would be an important nutrient if all it did was help reduce obesity. But as we've seen, chromium helps insulin serve many important functions. As a result, chromium helps people achieve a number of health-related goals, from controlling diabetes to building muscle to slowing the clock of age.

Help for Diabetics

Chromium, especially in picolinate form, can help both those who wish to reduce their risk of developing diabetes and those who have already developed the disease. Diabetes, unless correctly controlled, is a forerunner of disease and early death.[6] One-fourth of all kidney-failure cases and one-half of all amputations have been attributed to diabetic complications. Diabetes is the leading cause of blindness in the United States, and diabetics have a high risk of developing poor circulation and heart disease. Atherosclerosis—the accumulation of fat on the blood vessel walls—is two to three times more common in diabetics than in nondiabetics. Diabetics run twice the risk of developing high blood pressure and six times the risk of having a stroke as nondiabetics.

Both insulin-dependent and non-insulin-dependent diabetics can derive great benefits from chromium picolinate. The former are often able to decrease the amount of insulin they must inject daily, while the latter are often able to either reduce the amount of drugs they must take or eliminate the drugs altogether. As we've stressed throughout this book, insulin is a vital substance, but too much can cause problems. When insulin, whether injected or produced naturally, is working effectively, the long-term effects of diabetes can be lessened or eliminated. Diabetics should think of chromium picolinate as an insulin booster. (See Chapter 7.)

Chromium in the Gym

Chromium reverses and prevents insulin impotence. As a result, muscle cells can use glucose and fat for energy and amino acids for protein-building. This certainly helps muscle development, but there is another factor involved. Like all cells, muscles must be stimulated by one of the androgen hormones before they will grow and develop. The hormone

DHEA can stimulate the development of muscle, either directly or by being converted into testosterone, the most potent androgen. But there is very little DHEA in the bloodstream when insulin isn't working efficiently. The elevated blood-glucose levels caused by insulin impotence lead to the secretion of more insulin into the blood. The increased insulin levels, in turn, cause less DHEA to be produced.

Our studies show that exercise, even without chromium supplements, improves insulin efficiency. That isn't surprising, since exercise burns glucose, and excess glucose can interfere with the ability of the cells to take in fuel. But the combination of exercise and chromium picolinate is especially beneficial in improving insulin efficiency. With less insulin in the bloodstream, much more DHEA is produced and more muscle is formed. Muscles pay a heavy price when insulin impotence develops, but a combination of exercise, moderate carbohydrate intake, and chromium picolinate can help you build muscles safely and naturally. (See Chapter 8.)

You're as Young as You Feel

Many factors are involved in how fast the body's cells age. Experiments with both animals and humans show that chromium and insulin are among those factors.

Insulin impotence can cause elevated levels of LDL cholesterol in the bloodstream. This cholesterol is attacked by oxidants, causing it to accumulate in cells that line the blood vessels. This not only leads to the increased chance of a clot becoming stuck in the narrowed vessels, but also slows the flow of blood, which robs the cells of life-giving oxygen. Chromium picolinate supplements help keep LDL cholesterol from being produced. Insulin impotence also leads to higher levels of glucose in the bloodstream, glucose that can end up attaching itself to various proteins and causing vari-

ous problems. Chromium picolinate can reduce protein glycation.

The hormone DHEA has been shown to have several
anti-aging properties. DHEA production decreases when
insulin isn't effective, and insulin impotence increases with
age. For example, lower DHEA levels can lead to
osteoporosis, a weakening of the bones that often occurs in
older women.

Because insulin impotence alters so many bodily functions,
a combination of factors linked to insulin probably leads to
cell degeneration. And insulin needs chromium to be efficient. Chromium helps cells respond to insulin quickly and
effectively, thus improving bodily function. Therefore, chromium picolinate helps the body ward off disease and stay
vigorous. (See Chapter 9.)

CHROMIUM PICOLINATE AND SAFETY

Like all food supplements, chromium picolinate has been
thoroughly tested to determine if it is safe for human consumption. So far, all research shows a large margin of safety
between the effective dose and the toxic dose. After chromium picolinate is absorbed and used, any excess chromium
is excreted through the urine. Thus, chromium from food and
food supplements does not accumulate in the body. Picolinate is transferred to the liver, where it is coupled with a
chemical that inactivates and detoxifies it. It too is then excreted through the urine.

Several years ago, Dr. Bruce Ames at the University of
California—Berkeley devised a very effective method for
testing chemicals that might cause cancer. In this method,
known as the Ames test, the chemical in question is added to
five different kinds of bacteria. Bacteria reproduce very rapidly, so the addition of a cancer-causing chemical to their
food shows very obvious changes in their offspring. When

this happens, the chemical is said to test positive, and must be subjected to further tests.

Biodevelopment Laboratories, an independent laboratory in Cambridge, Massachusetts, used the Ames test on chromium picolinate. It tested negative in this highly reliable and widely used test.[7]

Dr. Richard Anderson and his colleagues at the United States Department of Agriculture also did a study that demonstrates the safety of chromium picolinate.[8] Ten different groups of rats were fed five different amounts of either chromium salts or chromium picolinate for twenty-four weeks, or about eight years in a human life. The highest level of chromium used in the study was equivalent to an adult human eating 2 grams—2 million micrograms—of chromium daily for eight to ten years, compared with my recommended daily chromium intake of between 400 and 600 micrograms. The livers and kidneys of these rats showed no signs of toxicity, even at the highest level of chromium intake.

A researcher in Japan studied the long-term, or chronic, effect of consuming picolinate. Using rats, he found the chronic toxicity level to be equivalent to a human adult consuming 52.5 grams—about 2 ounces—of picolinate a day.[9] A person taking a supplement with as much as 800 micrograms of chromium in picolinate form would absorb less than 7.0 milligrams of picolinate. That's seven one-thousandths of a gram, or less than one-thousandth of an ounce.

When I and my colleagues studied picolinate excretion, we found about 20 milligrams of picolinate in the urine of people who weren't even taking a supplement with any form of picolinate in it.[10] This means that the body makes about three times more picolinate than a person taking a big supplement of chromium picolinate would get. Keep in mind that the body does make picolinate, but not in the amounts that are needed to help all the essential minerals, including chromium, pass through the walls of the intestines and into the bloodstream.

The longevity studies with rats and the studies conducted by Dr. Merlin Lindemann with pregnant pigs (see Chapter 3) also provide very good evidence of chromium picolinate's safety. In one study, Dr. Lindemann fed the pigs the equivalent of 1,000 micrograms of chromium a day in humans. The pigs remained healthy and produced more and better litters.[11] In the longevity studies, the rats were fed the equivalent of about 625 micrograms of chromium a day in humans, and that only made them live longer.[12]

A Contrary Opinion

As a research scientist, my position in evaluating scientific research must always be unbiased—that is, even though I may see chromium picolinate as an important new find, I must be open to the work of others. There has been a study that has shown chromium picolinate to have harmful effects. However, after careful review, many scientists, including myself, believe that study to have been flawed.

In the spring of 1995, scientists at Dartmouth College and George Washington University first reported that chromium picolinate caused damage in hamster ovary cells being grown in test tubes.[13] Later, the study was published in a scientific journal.[14]

It was difficult to determine exactly how much chromium picolinate was used in this study. The abstract, or summary, contained no information on how much chromium picolinate was used. When first contacted, the authors explained how they did the study, and it turned out that they used 30,000 times the amount of chromium found in people who take chromium picolinate. When other scientists were able to see the data at a meeting of biologists in Atlanta, they were told that chromosome damage resulted when the study's authors used 6,000 times as much chromium as is found in people who take supplements. When the study was published, it said that the amount used to produce chromosome damage was 3,000 times as much as would be found in a person taking 200 micrograms a day of chromium in the form of picolinate.

In the paper that was published, the authors made some incorrect assumptions, and tried to leave the reader with the impression that the amount of chromium picolinate used was really not much higher than what would be found in humans given a chromium picolinate supplement. Although the authors did their experiments with cells in test tubes, their assumptions and conclusions were based on observations made on living animals. Despite the claims based on these questionable test-tube experiments, there is no evidence that chromium picolinate is toxic when eaten.

Possible Side Effects

Despite chromium picolinate's safety, reports of a few adverse reactions have surfaced. Some people seem to develop a slight skin rash or dizziness when they start taking chromium picolinate supplements. The reason for this is not perfectly clear, but these conditions should not be ignored.

These symptoms could mean that a person has extremely sensitive insulin, or that he or she is one of those rare individuals who already has enough chromium and doesn't need a supplement. More likely, though, is the possibility that the person has blood-insulin levels that are too high, so that an influx of chromium would cause the amount of blood sugar to drop too quickly. If such symptoms do occur, I suggest cutting the dosage in half. If symptoms persist, stop altogether and see your doctor.

Persistent dizziness that occurs an hour or two after taking chromium picolinate may be a sign of a more serious condition. If this happens to you and you are not taking insulin or glucose-control drugs, you may have true hypoglycemia. Unlike the reactive hypoglycemia we discussed in Chapter 7, true hypoglycemia occurs when the body cannot make the hormone needed to keep blood-glucose levels from dropping too low. If the dizziness still occurs even after you've reduced the dosage, stop taking the supplements and see your doctor.

Diabetics who start taking chromium picolinate need to monitor their blood sugar levels very, very carefully (see Chapter 7). Chromium picolinate increases the efficiency of both injected insulin and oral drugs. This could lead to hypoglycemic shock, in which blood-glucose levels drop too low. If you are a diabetic, see your doctor before using chromium picolinate.

HELP WILL HAVE TO COME FROM SUPPLEMENTS

Chromium is a vital mineral, but you aren't going to get it from your food.[15] No particular food group can be singled out as a good source of chromium. Dairy products generally have less than 1 microgram of chromium a serving. Meats, poultry, and fish are a little better, with 1 to 10 micrograms a serving. Grain products, including whole wheat bread, don't provide more than 8 micrograms a serving. Fruits and vegetables, generally good sources of nutrients, don't have much more than a few micrograms of chromium a serving. The one exception in this food group is the old reliable, broccoli, with over 20 micrograms in a serving. Beer, made from chromium-rich brewer's yeast, doesn't have much more than 50 micrograms in a one-liter pitcher. Even the dietitians and nutritionists who constantly urge people to get nutrients from food sources rather than from supplements wouldn't recommend four to six pitchers of beer a day to meet your chromium requirements.

Dietary analyses show that there are, on average, about 14 micrograms of chromium in every 1,000 calories consumed.

To get enough chromium from the diet, one would have to eat over 14,000 calories a day! That's why people need to take chromium picolinate supplements. This is the only way we can rid the world of the epidemic caused by lack of chromium.

Many experiments show that the form of chromium used in a supplement determines whether or not someone who takes the supplement will derive benefits from it.[16–18] Chromium's presence alone in a food or supplement doesn't guarantee that the chromium will be used by the body. Many chemical forms of chromium have been tested and discarded as useless. Several chemical forms of chromium are currently being marketed even though they don't effectively prevent insulin impotence. As we've seen throughout this book, chromium picolinate is the most effective way to give the body chromium in a form it can actually use.

Keep in mind that not all chromium picolinate is created equal. The chromium picolinate that has been proven effective in all the studies we've discussed has been manufactured to strict standards regarding quality and dosage. But as is the case with many other nutritional supplements, there are copycat versions on the market. These versions offer no guarantee of quality control. Therefore, you should look for U.S. Patent Number 4,315,927, Reissue Number 33,988—licensed to Nutrition 21, San Diego, California—when looking for chromium picolinate. If the product you buy doesn't have one of these numbers, it may be completely ineffective.

HOW MUCH DO YOU NEED?

Estimates of chromium requirements are based on the research done with chromium picolinate in both humans and animals. These recommendations assume you will be using chromium picolinate, but the quantity indicated is for chromium only, since that's the way chromium supplements are sold.

Children need the following amounts of chromium a day:

- Up to age one—50 micrograms
- Ages one to six—100 micrograms
- Ages seven to ten—200 micrograms

Teenaged boys (ages eleven to nineteen) should take from 300 to 400 micrograms of chromium a day, while teenaged girls need between 200 and 300 micrograms. Girls require less chromium because of their lower calorie requirements.

All adults up to about age sixty-five would be well advised to use a daily supplement of 400 micrograms. Elderly people should take at least 200 micrograms a day. Dieters, diabetics, and bodybuilders, as well as those trying to lower their cholesterol levels or increase their calcium levels, may need up to 600 micrograms a day (see Chapters 6, 7, 8, or 9).

Of course, these are general suggestions. They are based on estimations of calorie intake. If you are very active, or if you do very little physical exercise, you may want to adjust your chromium intake to coincide more closely with the number of calories you consume in a day. Close scrutiny of all the studies with both humans and animals shows the most dramatic and consistent results were obtained with at least 200 micrograms of chromium per every 1,000 calories.

CHROMIUM FOR LIFE

By using a daily chromium supplement, ideally as chromium picolinate, along with a general vitamin and mineral supplement, you should have less trouble in controlling body composition, maintaining optimal health, and retaining youth. When all systems are working correctly, your body puts nutrients where they belong, and uses them only when they are needed and only for their intended purposes.

To begin with, you will eat only when your body needs to increase its fuel supply because your brain's fuel gauge will be working properly. You won't crave sugar-loaded goodies when your cells think they need a quick fuel fix.

A little of the glucose from the food you eat will be used to help make fat if it's needed, and just the right amount of glucose will stay in the bloodstream to fuel the nerve cells. The fat from the diet will be shunted into fat cells, but this fat will be readily available for use by the muscle cells. The muscle cells will have an ample supply of high-quality fuel, so they won't use up their stored glucose and they won't have to destroy their own proteins to get the fuel they need. This will allow the muscles to grow and develop.

There will be less glucose floating around in the bloodstream. This means there will be less glucose attaching itself to various proteins. Less glycated protein will help prevent vascular disease and thus help promote a healthier, perhaps even longer, life.

Everybody needs chromium—young, old, healthy, bedridden, thin, obese. Don't put off taking chromium supplements until you contract a deficiency disease. Chromium is a nutrient for life.

There's a better body inside you. Find it and use it.

Good health to you and your loved ones.

NOTES

Introduction

1. Ravina A and Slezack L. The clinical use of the trace element chromium (III) in the treatment of diabetes mellitus. *Harefuah* 125:142–148, 1993.
2. Jovanovic-Peterson L, Gutierrez M, and Peterson CM. Chromium supplementation for gestational diabetic women improves glucose tolerance and decreases hyperinsulinemia. *Journal of the American College of Nutrition* 14:530, 1995.
3. Press RI, Geller J, and Evans GW. The effect of chromium picolinate on serum cholesterol and apolipoprotein fractions in human subjects. *Western Journal of Medicine* 152:41–45, 1990.
4. Evans GW. Effect of chromium picolinate on insulin controlled parameters in humans. *International Journal of Biosocial and Medical Research* 11:163–180, 1989.
5. Evans GW. Effect of chromium picolinate on insulin controlled parameters in humans. *International Journal of Biosocial and Medical Research* 11:163–180, 1989.
6. Hasten DL, Rome EP, Franks BD, and Hegsted M. Effects of chromium picolinate on beginning weight training students. *International Journal of Sport Nutrition* 2:343–350, 1992.
7. Hallmark MA, Reynolds TH, DeSouza CA, Dotson CO, Anderson RA, and Rogers MA. Effects of chromium supplementation and resis-

tive training on muscle strength and lean body mass in untrained men. *Medicine and Science in Sports and Exercise* 25:S101, 1993.

8. Kaats GR, Blum K, Fisher JA, and Adelman JA. The effects of chromium picolinate suppplementation on body composition: a randomized double blind placebo controlled study. *Current Therapeutic Research*, February 1996 (at press).

9. Bahadori B, Habersack S, Schneider H, Wascher TC, and Toplak H. Treatment with chromium picolinate improves lean body mass in patients following weight reduction. *International Journal of Obesity* 19(Suppl 2):38, 1995.

10. Evans GW. The effect of exercise and chromium picolinate on dehydroepiandrosterone (DHEA) in males. Bemidji State University, 1992.

11. Evans GW, Swenson G, and Walters K. Chromium picolinate decreases calcium excretion and increases dehydroepiandrosterone (DHEA) in post menopausal women. *FASEB Journal* 9:525, 1995.

Chapter 1
What Is Chromium Picolinate?

1. Freund H, Atamian S, and Fischer JE. Chromium deficiency during total parenteral nutrition. *Journal of the American Medical Association* 214:496–498, 1979.

2. Jeejeebhoy KN, Chu RC, Marliss EB, Greenberg R, and Bruce-Robertson AS. Chromium deficiency, glucose intolerance and neuropathy reversed by chromium supplementation in a patient receiving long-term total parenteral nutrition. *American Journal of Clinical Nutrition* 30:531–538, 1977.

3. Mertz W. Chromium occurrence and function in biological systems. *Physiological Reviews* 49:163–203, 1969.

4. Pi-Sunyer FX and Offenbacher EG. Chromium. *Nutrition Reviews' Present Knowledge in Nutrition* 5:571–586, 1984.

5. Rabinowitz MB, Gonick HC, Levin SR, and Davidson MB. Clinical trial of chromium and yeast supplements on carbohydrate and lipid metabolism in diabetic men. *Biological Trace Element Research* 5:449–466, 1983.

6. Schwarz K and Mertz W. Chromium (III) and the glucose tolerance factor. *Archives of Biochemistry and Biophysics* 85:292–295, 1959.

7. Anderson RA and Kozlovsky AS. Chromium intake, absorption and excretion of subjects consuming self-se-

lected diets. *American Journal of Clinical Nutrition* 41:1177–1183, 1985.

8. Anderson RA. Chromium metabolism and its role in disease processes in man. *Clinical Physiology and Biochemistry* 4:31–41, 1986.

9. Rawn JD. *Biochemistry.* Neil Patterson Publishers: Burlington, NC. 1989.

10. Schauf CL, Moffett DF, and Moffett SB. *Human Physiology.* Times Mirror Mosby College Publishing: St. Louis, MO. 1990.

11. Stryer L. *Biochemistry.* Third edition. W. H. Freeman: New York. 1988.

12. Whitney EN and Hamilton EMN. *Understanding Nutrition.* West Publishing: St. Paul, MN. 1984.

13. Evans GW, Johnson PE, Brushmiller JG, and Ames RW. Detection of labile zinc-binding ligands in biological fluids by modified gel filtration chromatography. *Analytical Chemistry* 51:839–843, 1979.

14. Evans GW and Johnson PE. Characterization and quantitation of a zinc-binding ligand in human milk. *Pediatric Research* 14:876–880, 1980.

15. Evans GW. The role of picolinate in metal metabolism. *Life Chemistry Reports* 1:57–67, 1982.

16. Krieger I, Cash R, and Evans GW. Picolinate in acrodermatitis enteropathica: evidence for a disorder of tryptophan metabolism. *Journal of Pediatric Gastroenterology and Nutrition* 3:62–68, 1984.

17. Krieger I and Evans GW. A variant of acrodermatitis enteropathica without hypozincemia. Therapeutic effect of a pancreatic enzyme preparation due to a zinc binding ligand. *Journal of Pediatrics* 96:32–35, 1980.

18. Barrie SA, Wright JV, Pizzorno JE, Kutter E, and Barron PC. Comparative absorption of zinc picolinate, zinc citrate and zinc gluconate in humans. *Agents and Actions* 5:1–6, 1986.

19. Boosalis MG, Evans GW, and McClain CJ. Impaired handling of orally administered zinc in pancreatic insufficiency. *American Journal of Clinical Nutrition* 37:268–271, 1983.

20. Evans GW. Normal and abnormal zinc absorption in man and animals: the tryptophan connection. *Nutrition Reviews* 38:137–141, 1980.

21. Evans GW. *The Picolinates.* Keats Publishing: New Caanan, CT. 1989.

22. Evans GW and Johnson EC. Effect of iron, vitamin B-6 and picolinate on zinc absorption in the rat. *Journal of Nutrition* 111:68–75, 1981.

23. Evans GW and Johnson EC. Growth stimulating effect of picolinate added to rat diets. *Proceedings of the Society for Experimental Biology and Medicine* 165:457–461, 1980.

24. Evans GW and Johnson EC. Zinc absorption in rats fed a low protein diet and a low protein diet supplemented with tryptophan or picolinate. *Journal of Nutrition* 110:1076–1080, 1980.

25. Evans GW and Johnson EC. Zinc concentrations of liver and kidneys from rat pups nursing dams fed supplemental zinc dipicolinate or zinc acetate. *Journal of Nutrition* 110:2121–2124, 1980.

26. Johnson PE, Evans GW, and Hunt JR. The effect of picolinate supplementation on zinc absorption by men fed a low tryptophan diet. *Nutrition Research* 8:119–127, 1988.

27. Krieger I and Statter M. Tryptophan deficiency and picolinate: effect on zinc metabolism and clinical manifestations of pellagra. *American Journal of Clinical Nutrition* 46:511–517, 1987.

28. Seal CJ and Heaton FW. Chemical factors affecting the intestinal absorption of zinc in vitro and in vivo. *British Journal of Nutrition* 50:317–324, 1983.

29. Seri S, Aquilio E, Continenza A, and Ricciardi G. Effects of dietary tryptophan bioavailability on zinc absorption in rats. *IRCS Medical Science* 12:452–453, 1984.

30. Schroeder HA, Nason AP, and Tipton IH. Chromium deficiency as a factor in atherosclerosis. *Journal of Chronic Diseases* 23:123–142, 1970.

31. Schroeder HA. Serum cholesterol and glucose levels in rats fed refined and less refined sugars and chromium. *Journal of Nutrition* 97:237–242, 1969.

32. Schroeder HA. The role of chromium in mammalian nutrition. *American Journal of Clinical Nutrition* 21:230–244, 1968.

33. Schroeder HA. *The Poisons Around Us.* Keats Publishing: New Canaan, CT. 1974.

34. Hunt JV, Smith CCT, and Wolff SP. Autoxidative glycosylation and possible involvement of peroxides and free radicals in LDL modification by glucose. *Diabetes* 39:1420–1424, 1990.

35. Rawn JD. *Biochemistry.* Neil Patterson Publishers: Burlington, NC. 1989.

36. Schauf CL, Moffett DF, and Moffett SB. *Human Physiology.* Times Mirror Mosby College Publishing: St. Louis, MO. 1990.

37. Evans GW and Press RI. Cholesterol and glucose low-

ering effect of chromium picolinate. *FASEB Journal* 3:A761, 1989.

38. Press RI, Geller J, and Evans GW. The effect of chromium picolinate on serum cholesterol and apolipoprotein fractions in human subjects. *Western Journal of Medicine* 152:41–45, 1990.

39. Jeejeebhoy KN, Chu RC, Marliss EB, Greenberg R, and Bruce-Robertson AS. Chromium deficiency, glucose intolerance and neuropathy reversed by chromium supplementation in a patient receiving long-term total parenteral nutrition. *American Journal of Clinical Nutrition* 30:531–538, 1977.

40. Evans GW. Effect of chromium picolinate on insulin controlled parameters in humans. *International Journal of Biosocial and Medical Research* 11:163–180, 1989.

41. Evans GW. An inexpensive, convenient adjunct for the treatment of diabetes. *Western Journal of Medicine* 155:549, 1991.

42. Evans GW. Effect of chromium picolinate on insulin controlled parameters in humans. *International Journal of Biosocial and Medical Research* 11:163–180, 1989.

43. Nestler JE, Clore JN, and Blackard WG. Dehydroepiandrosterone: the "missing link" between hyperinsulinemia and atherosclerosis? *FASEB Journal* 6:3073–3075, 1992.

44. Evans GW, Swenson G, and Walters K. Chromium picolinate decreases calcium excretion and increases dehydroepiandrosterone (DHEA) in post menopausal women. *FASEB Journal* 9: 525, 1995.

Chapter 2
Chromium Picolinate Is for Real—Human Studies

1. Gordon JB. An easy and inexpensive way to lower cholesterol? *Western Journal of Medicine* 154:352, 1991.

2. Mehler AH. Formation of picolinic and quinolinic acids following enzymatic oxidation of 3-hydroxyanthranilic acid. *Journal of Biological Chemistry* 218:241–254, 1956.

3. Lee NA and Reasner CA. Beneficial effect of chromium supplementation on serum triglyceride levels in NIDDM. *Diabetes Care* 17:1449–1452, 1994.

4. Hasten DL, Rome EP, Franks BD, and Hegsted M. Effects of chromium picolinate on beginning weight training students. *International Journal of Sport Nutrition* 2:343–350, 1992.

5. Evans GW. Effect of chromium picolinate on insulin

controlled parameters in humans. *International Journal of Biosocial and Medical Research* 11:163–180, 1989.

6. Colgan M. *Optimum Sports Nutrition*. Advanced Research Press: Ronkonkoma, NY. 1993.

7. Kaats GR, Blum K, Fisher JA, and Adelman JA. The effects of chromium picolinate suppplementation on body composition: a randomized double blind placebo controlled study. *Current Therapeutic Research*, February 1996 (at press).

8. Kaats GR, Wise JA, Blum K, Morin RJ, Adelman JA, Craig J, and Croft HA. The short-term therapeutic efficacy of treating obesity with a plan of improved nutrition and moderate calorie restriction. *Current Therapeutic Research* 51:261–274, 1992.

9. Bahadori B, Habersack S, Schneider H, Wascher TC, and Toplak H. Treatment with chromium picolinate improves lean body mass in patients following weight reduction. *International Journal of Obesity* 19(Suppl 2):38, 1995.

10. Ravina A and Slezack L. The clinical use of the trace element chromium (III) in the treatment of diabetes mellitus. *Harefuah* 125:142–148, 1993.

11. Jovanovic-Peterson L, Gutierrez M, and Peterson CM. Chromium supplementation for gestational diabetic women improves glucose tolerance and decreases hyperinsulinemia. *Journal of the American College of Nutrition* 14:530, 1995.

Chapter 3
Pigs Don't Lie—Animal Studies

1. Bergen WG and Merkel RA. Body composition of animals treated with partitioning agents: implications for human health. *FASEB Journal* 5:2951–2957, 1991.

2. Page TG. Chromium, tryptophan, and picolinate in diets for pigs and poultry. PH.D. diss., Louisiana State University, 1991, 1–92.

3. Page TG, Boleman SL, Pike MM, and Southern LL. Effect of chromium picolinate on carcass traits and aging on pork quality. *Journal of Animal Science* 70(Suppl 1):322, 1992.

4. Page TG, Southern LL, Herbert JA, Ward TL, and Achee VN. Effect of chromium picolinate on serum cholesterol, egg production, egg cholesterol and egg quality of laying hens. *Poultry Science* 70(Suppl l):91, 1991.

5. Page TG, Southern LL, Ward TL, and Thompson DL Jr. Ef-

fect of chromium picolinate on growth and serum and carcass traits of growing-finishing pigs. *Journal of Animal Science* 71:656–662, 1993.

6. Page TG, Southern LL, and Ward TL. Effect of chromium picolinate on growth, serum and carcass traits, and organ weights of growing-finishing pigs from different ancestral sources. *Journal of Animal Science* 70(Suppl 1):389, 1992.

7. Page TG, Ward TL, and Southern LL. Chromium supplementation of corn-soybean meal diets for finishing swine. *Journal of Animal Science* 68(Suppl 1):39, 1990.

8. Page TG. Chromium, tryptophan, and picolinate in diets for pigs and poultry. PH.D. diss., Louisiana State University, 1991, 1–92.

9. Page TG, Southern LL, Ward TL, and Thompson DL Jr. Effect of chromium picolinate on growth and serum and carcass traits of growing-finishing pigs. *Journal of Animal Science* 71:656–662, 1993.

10. Evock-Clover CM, Anderson RA, and Steele NC. Dietary chromium supplementation with or without somatotropin treatment alters serum hormones and metabolites in growing pigs without affecting growth performance. *Journal of Nutrition* 123:1504–1512, 1993.

11. Mooney KW and Cromwell GL. Effect of chromium picolinate on performance, carcass composition and tissue accretion in growing-finishing pigs. *Journal of Animal Science* 71(Supp 1):167, 1993.

12. Hu CY, Oregon State University, letter to author, 19 June 1991.

13. Liarn TF, Chen SY, Chen CL, and Hu CY. The effects of various levels of chromium picolinate on growth performance and serum traits of pigs. *Journal of the Chinese Society of Animal Science* 22(4):349–357, 1993.

14. Lindemann MD, Harper AP, and Kornegay ET. Reproductive response in swine to the supplementation of chromium from chromium picolinate. *Journal of Animal Science* 72(Supp 1):66, 1994.

15. Lindemann MD, Wood CM, Harper AF, and Kornegay ET. Chromium picolinate addition to diets of growing-finishing pigs. *Journal of Animal Science* 71(Suppl 1):14, 1993.

16. Lindemann MD, Wood CM, Harper AF, Kornegay ET, and Anderson RA. Dietary chromium picolinate additions improve gain: feed and carcass characteristics in growing-finishing pigs and increase litter size in reproducing sows. *Journal of Animal Science* 73:457–465, 1995.

17. Morris, OS, Guidry KA, Hegsted M, and Hasten D. Effects of dietary chromium supplementation on cardiac mass, metabolic enzymes, and contractile proteins. *Nutrition Research* 15:1045–1052, 1995.

18. Hasten DL, Hegsted M, and Glickman-Weiss EL. Effects of chromium picolinate on the body composition of the rat. *FASEB Journal* 7:A77, 1993.

19. Hasten D, Siver F, Fornea S, Savoie S, and Hegsted M. Dosage effects of chromium picolinate on body composition. *FASEB Journal* 8:A194, 1994.

20. Cerami A. Hypothesis: Glucose as a mediator of aging. *Journal of the American Geriatric Society* 33:626–634, 1985.

21. Masoro EJ, Katz MS, and McMahan CA. Evidence for the glycation hypothesis of aging from the food-restricted rodent model. *Journal of Gerontology: Biological Science* 44:B20–B22, 1989.

22. McCay C, Crowell M, and Maynard L. The effect of retarded growth upon the life and upon the ultimate size. *Journal of Nutrition* 10:63–79, 1935.

23. Masoro EJ, Yu BP, and Bertrand HA. Action of food restriction in delaying the aging process. *Proceedings of the National Academy of Sciences, USA* 79:4239–4241, 1982.

24. Evans GW. Effect of chromium picolinate on insulin controlled parameters in humans. *International Journal of Biosocial and Medical Research* 11:163–180, 1989.

25. Schroeder HA. The role of chromium in mammalian nutrition. *American Journal of Clinical Nutrition* 21:230–244, 1968.

26. Evans GW and Meyer LK. Life span is increased in rats supplemented with a chromium-pyridine 2 carboxylate complex. *Advances in Scientific Research* 1:19–23, 1994.

27. Page TG. Chromium, tryptophan, and picolinate in diets for pigs and poultry. PH.D. diss., Louisiana State University, 1991, 1–92.

28. Kitchalong L, Fernandez JM, Bunting LD, Chapa AM, Sticker LS, Amoikon EK, Ward TL, Bidner TD, and Southern LL. Chromium picolinate supplementation in lamb rations. Effects on performance, nitrogen balance, endocrine and metabolic parameters. *Journal of Animal Science* 71(Supp 1):291, 1993.

29. Fernandez JM, Bunting LD, Thompson DL Jr, and Southern LL. Glucose tolerance, insulin sensitivity, and growth hormone levels in steer and heifer calves fed chromium

picolinate. *FASEB Journal* 7:A525, 1993.

30. Gurney HC Jr. Muscle atrophy—the chromium connection. *Journal of the American Holistic Veterinary Medical Association* 9:5–6, 1990.

31. Liarn TF, Chen SY, Shiau SP, Froman DP, and Hu CY. Chromium picolinate reduces serum and egg cholesterol of laying hen. *Professional Animal Science*, March 1996 (at press).

32. Page TG. Chromium, tryptophan, and picolinate in diets for pigs and poultry. PH.D. diss., Louisiana State University, 1991, 1–92.

33. Page TG, Southern LL, Herbert JA, Ward TL, and Achee VN. Effect of chromium picolinate on serum cholesterol, egg production, egg cholesterol and egg quality of laying hens. *Poultry Science* 70(Suppl 1):91, 1991.

34. To figure Chro Cals, divide the amount of chromium in a certain amount of food by the number of Calories, and then multiply by 1,000 to obtain a whole number. For example, if 1 kilogram of pig feed contains 200 micrograms of chromium and 3,000 Calories, the diet has 0.066 micrograms chromium per Calorie, or 66 micrograms chromium per 1,000 Calories, or 66 Chro Cals.

Chapter 4
Why Your Body Needs Chromium Picolinate

1. Anderson RA and Kozlovsky AS. Chromium intake, absorption and excretion of subjects consuming self-selected diets. *American Journal of Clinical Nutrition* 41:1177–1183, 1985.

2. Schwarz K and Mertz W. A glucose tolerance factor and its differentiation from factor 3. *Archives of Biochemistry and Biophysics* 72:515–518, 1957.

3. Schwarz K and Mertz W. Chromium (III) and the glucose tolerance factor. *Archives of Biochemistry and Biophysics* 85:292–295, 1959.

4. Schroeder HA, Nason AP, and Tipton IH. Chromium deficiency as a factor in atherosclerosis. *Journal of Chronic Diseases* 23:123–142, 1970.

5. Schroeder HA. Serum cholesterol and glucose levels in rats fed refined and less refined sugars and chromium. *Journal of Nutrition* 97:237–242, 1969.

6. Schroeder HA. The role of chromium in mammalian nutrition. *American Journal of Clinical Nutrition* 21:230–244, 1968.

7. Mertz W. Chromium occurrence and function in biological systems. *Physiological Reviews* 49:163–203, 1969.

8. Jeejeebhoy KN, Chu RC, Marliss EB, Greenberg R, and Bruce-Robertson AS. Chromium deficiency, glucose intolerance and neuropathy reversed by chromium supplementation in a patient receiving long-term total parenteral nutrition. *American Journal of Clinical Nutrition* 30:531–538, 1977.

9. Freund H, Atamian S, and Fischer JE. Chromium deficiency during total parenteral nutrition. *Journal of the American Medical Association* 214:496–498, 1979.

10. Brown RO, Forloines-Lynn S, Cross RE, and Heizer WD. Chromium deficiency after long-term parenteral nutrition. *Digestive Disease Science* 31:661–664, 1986.

11. Editorial. Is chromium essential for humans? *Nutrition Reviews* 46:17–20. 1988.

12. Mertz W. The essential trace elements. *Science* 213:1332–1338, 1981.

13. Pi-Sunyer FX and Offenbacher EG. Chromium. *Nutrition Reviews' Present Knowledge in Nutrition* 5:571–586, 1984.

14. Editorial. Is chromium essential for humans? *Nutrition Reviews* 46:17–20. 1988.

15. Mertz W. The essential trace elements. *Science* 213:1332–1338, 1981.

16. Anderson RA. Chromium metabolism and its role in disease processes in man. *Clinical Physiology and Biochemistry* 4:31–41, 1986.

17. Pi-Sunyer FX and Offenbacher EG. Chromium. *Nutrition Reviews' Present Knowledge in Nutrition* 5:571–586, 1984.

18. Offenbacher EG and Pi-Sunyer FX. Beneficial effects of chromium-rich yeast on glucose tolerance and blood lipids in elderly subjects. *Diabetes* 29:919–925, 1980.

19. Rabinowitz MB, Gonick HC, Levin SR, and Davidson MB. Clinical trial of chromium and yeast supplements on carbohydrate and lipid metabolism in diabetic men. *Biological Trace Element Resarch* 5:449–466, 1983.

20. Toepfer EW, Mertz W, Polansky MM, Roginski EE, and Wolf WR. Synthetic organic chromium complexes and glucose tolerance. *Journal of Agriculture and Food Science* 25:162–165, 1977.

21. Anderson RA, Bryden NA, and Polansky MM. Dietary chromium intake. Freely chosen diets, institutional diets and individual foods. *Biological Trace Element Research* 32:117–121, 1992.

22. Mehler AH. Formation of picolinic and quinolinic acids following enzymatic oxidation of 3-hydroxyanthranilic

acid. *Journal of Biological Chemistry* 218:241–254, 1956.

23. Moynahan EJ. Acrodermatitis enteropathica: a lethal inherited human zinc deficiency disorder. *Lancet* 2:399–400, 1974.

24. Evans GW. The role of picolinate in metal metabolism. *Life Chemistry Reports* 1:57–67, 1982.

25. Evans GW and Johnson PE. Characterization and quantitation of a zinc-binding ligand in human milk. *Pediatric Research* 14:876–880, 1980.

26. Evans GW. The role of picolinate in metal metabolism. *Life Chemistry Reports* 1:57–67, 1982.

27. Evans GW. Normal and abnormal zinc absorption in man and animals: the tryptophan connection. *Nutrition Reviews* 38:137–141, 1980.

28. Evans GW. *The Picolinates.* Keats Publishing: New Caanan, CT. 1989.

29. Krieger I and Evans GW. A variant of acrodermatitis enteropathica without hypozincemia. Therapeutic effect of a pancreatic enzyme preparation due to a zinc binding ligand. *Journal of Pediatrics* 96:32–35, 1980.

30. Krieger I, Cash R, and Evans GW. Picolinate in acrodermatitis enteropathica: evidence for a disorder of tryptophan metabolism. *Journal*

of Pediatric Gastroenterology and Nutrition 3:62–68, 1984.

Chapter 5
Chromium and Insulin— A Healthy Connection

1. Evans GW and Bowman TD. Chromum picolinate increases membrane fluidity and rate of insulin internalization. *Journal of Inorganic Biochemistry* 46:243–250, 1992.

2. Evock-Clover CM, Anderson RA, and Steele NC. Dietary chromium supplementation with or without somatotropin treatment alters serum hormones and metabolites in growing pigs without affecting growth performance. *Journal of Nutrition* 123:1504–1512, 1993.

3. Fernandez JM, Bunting LD, Thompson DL Jr, and Southern LL. Glucose tolerance, insulin sensitivity, and growth hormone levels in steer and heifer calves fed chromium picolinate. *FASEB Journal* 7:A525, 1993.

4. Freund H, Atamian S, and Fischer JE. Chromium deficiency during total parenteral nutrition. *Journal of the American Medical Association* 214:496–498, 1979.

5. Jeejeebhoy KN, Chu RC, Marliss EB, Greenberg R, and Bruce-Robertson AS. Chro-

mium deficiency, glucose intolerance and neuropathy reversed by chromium supplementation in a patient receiving long-term total parenteral nutrition. *American Journal of Clinical Nutrition* 30:531–538, 1977.

6. Kitchalong L, Fernandez JM, Bunting LD, Chapa AM, Sticker LS, Amoikon EK, Ward TL, Bidner TD, and Southern LL. Chromium picolinate supplementation in lamb rations. Effects on performance, nitrogen balance, endocrine and metabolic parameters. *Journal of Animal Science* 71(Supp 1):291, 1993.

7. Lindemann MD, Harper AP, and Kornegay ET. Reproductive response in swine to the supplementation of chromium from chromium picolinate. *Journal of Animal Science* 72(Supp 1):66, 1994.

8. Lindemann MD, Wood CM, Harper AF, Kornegay ET, and Anderson, RA. Dietary chromium picolinate additions improve gain: feed and carcass characteristics in growing-finishing pigs and increase litter size in reproducing sows. *Journal of Animal Science* 73:457–465, 1995.

9. Lindemann MD, Wood CM, Harper AF, and Kornegay ET. Chromium picolinate addition to diets of growing-finishing pigs. *Journal of Ani-*

mal Science 71(Suppl 1):14, 1993.

10. Mertz W. Chromium occurrence and function in biological systems. *Physiological Reviews* 49:163–203, 1969.

11. Ravina A and Slezack L. The clinical use of the trace element chromium (III) in the treatment of diabetes mellitus. *Harefuah* 125:142–148, 1993.

12. Steele NC, Richards MP, and Rosebrough RW. Effect of dietary chromium and protein status on hepatic insulin binding characteristics of swine. *Journal of Animal Science* 55(Suppl 1):300, 1982.

13. Ward TL, Berrio LF, Southern LL, Fernandez JM, and Thompson DL Jr. In-vivo and in-vitro evaluation of chromium tripicolinate on insulin binding in pig liver cell plasma membranes. *FASEB Journal* 8(4):A194, 1994.

14. Reaven GM. Role of insulin resistance in human disease. *Diabetes* 37:1595–1607, 1988.

15. Reaven GM. Insulin resistance, hyperinsulinemia, hypertryglyceridemia and hypertension: parallels between human disease and rodent models. *Diabetes Care* 14:195–202, 1991.

16. DeFronzo RA and Ferrannini E. Insulin resistance: a multifaceted syndrome responsible for NIDDM, obe-

sity, hypertension, dyslipi-
demia, and athersclerotic
vascular disease. *Diabetes
Care* 14:173–194, 1991.

17. Atkins R. *Dr. Atkins' New
Diet Revolution.* M. Evans:
New York. 1992.

18. Schroeder HA. The role of
chromium in mammalian
nutrition. *American Journal of
Clinical Nutrition* 21:230–244,
1968.

19. Marshall S, Garvey WT, and
Traxinger RR. New insights
into the metabolic regulation
of insulin action and insulin
resistance: role of glucose
and amino acids. *FASEB Journal* 5:3031–3036, 1991.

20. Klip A, Tsakiridis T, Marette
A, and Ortiz PA. Regulation
of expression of glucose
transporters by glucose: a review of studies in vivo and in
cell cultures. *FASEB Journal*
8:43–53, 1994.

21. Lienhard GE, Slot JW, James
DE, and Mueckler MM. How
cells absorb glucose. *Scientific
American* 226:86–91, 1992.

22. Kitchalong L, Fernandez JM,
Bunting LD, Chapa AM,
Sticker LS, Amoikon EK,
Ward TL, Bidner TD, and
Southern LL. Chromium pi-
colinate supplementation in
lamb rations. Effects on performance, nitrogen balance,
endocrine and metabolic parameters. *Journal of Animal
Science* 71(Supp 1):291, 1993.

23. Ward TL, Berrio LF, Southern LL, Fernandez JM, and
Thompson DL Jr. In-vivo and
in-vitro evaluation of chromium tripicolinate on insulin
binding in pig liver cell
plasma membranes. *FASEB
Journal* 8(4):A194, 1994.

24. Mertz W. Chromium occurrence and function in biological systems. *Physiological Reviews* 49:163–203, 1969.

25. Evans GW and Bowman TD.
Chromum picolinate in-
creases membrane fluidity
and rate of insulin internalization. *Journal of Inorganic Biochemistry* 46:243–250, 1992.

26. Anderson RA. Chromium
metabolism and its role in
disease processes in man.
*Clinical Physiology and Bio-
chemistry* 4:31–41, 1986.

27. Evans GW and Bowman
TD. Chromum picolinate increases membrane fluidi-
ty and rate of insulin internalization. *Journal of Inorganic Biochemistry* 46:243–
250, 1992.

28. Marshall S, Garvey WT, and
Traxinger RR. New insights
into the metabolic regulation
of insulin action and insulin
resistance: role of glucose
and amino acids. *FASEB Journal* 5:3031–3036, 1991.

29. Marshall S, Garvey WT, and
Traxinger RR. New insights
into the metabolic regulation
of insulin action and insulin

resistance: role of glucose and amino acids. *FASEB Journal* 5:3031–3036, 1991.

30. Klip A, Tsakiridis T, Marette A, and Ortiz PA. Regulation of expression of glucose transporters by glucose: a review of studies in vivo and in cell cultures. *FASEB Journal* 8:43–53, 1994.

31. Anderson RA, Polansky MM, Bryden NA, Patterson KY, Veillon C, and Glinsmann WH. Effect of chromium supplementation on urinary Cr excretion of human subjects and correlation of Cr excretion with selected clinical parameters. *Journal of Nutrition* 113:276–281, 1983.

32. Anderson RA, Polansky MM, and Bryden NA. Strenuous running: acute effects on chromium, copper, zinc and selected in urine and serum of male runners. *Biological Trace Element Research* 6:327–336, 1984.

33. Anderson RA. Chromium metabolism and its role in disease processes in man. *Clinical Physiology and Biochemistry* 4:31–41, 1986.

Chapter 6
Chromium Picolinate and Fat Loss

1. Kalimi M and Regelson W. *The Biologic Role of Dehydroepiandrosterone (DHEA).* Walter de Gruyter & Co.: Berlin. 1990.

2. Nestler JE, Clore JN, and Blackard WG. Dehydroepiandrosterone: the "missing link" between hyperinsulinemia and atherosclerosis? *FASEB Journal* 6:3073–3075, 1992.

3. Fox FW. The enigma of obesity. *Lancet* 2:1487–1488, 1973.

4. Astrup A. Thermogenesis in human brown adipose tissue and skeletal muscle induced by sympathomimetic stimulation. *Acta Endocrinology* 112(Suppl 278):8–30, 1986.

5. Campfield LA and Smith FJ. Modulation of insulin secretion by the autonomic nervous system. *Brain Research Bulletin* 5(Suppl 4):103–107, 1980.

6. Colquhoun EQ and Clark MG. Open question: has thermogenesis in muscle been overlooked and misinterpreted? *News In Physiological Sciences* 6:256–259, 1991.

7. Debons AF, Krimsky I, From A, and Cloutier RJ. Rapid effects of insulin on the hypothalamic satiety-center. *American Journal of Physiology* 217:1114–1118, 1969.

8. Lonnqvist F, Wennlund A, and Arner P. Antilipolytic effects of insulin and adenylate cyclase inhibitors on isolated

human fat cells. *International Journal of Obesity* 13:137–146, 1989.

9. Olivieri MC and Botellio LHP. Synergistic inhibition of hepatic ketogenesis in the presence of insulin and a cAMP antagonist. *Biochemistry and Biophysics Research Communications* 159:741–747, 1989.

10. Oomura Y and Kita H. Insulin acting as a modulator of feeding through the hypothalamus. *Diabetologia* 20: 290–298, 1981.

11. Rothwell NJ and Stock MJ. Insulin and thermogenesis. *International Journal of Obesity* 12:93–102, 1988.

12. Van Itallie TB. The glucostatic theory 1953–1988: roots and branches. *International Journal of Obesity* 14(Suppl 3):1–10, 1990.

13. Williamson DH. Role of insulin in the integration of lipid metabolism in mammalian tissues. *Biochemistry Society Transactions* 17:37–40, 1989.

14. McCarty MF. Hypothesis: sensitization of insulin-dependent hypothalamic glucoreceptors may account for the fat-reducing effects of chromium picolinate. *Journal of Optimal Nutrition* 2:36–53, 1993.

15. Whitney EN and Hamilton EMN. *Understanding Nutri-tion.* West Publishing: St. Paul, MN. 1984.

16. Pittman CS, Suda AK, Chambers JB Jr, and Ray GY. Impaired 3,5,3,'-triiodothyronine (T3) production in diabetic patients. *Metabolism* 28:333–338, 1979.

17. Astrup A. Thermogenesis in human brown adipose tissue and skeletal muscle induced by sympathomimetic stimulation. *Acta Endocrinology* 112(Suppl 278):8–30, 1986.

18. Colquhoun EQ and Clark MG. Open question: has thermogenesis in muscle been overlooked and misinterpreted? *News In Physiological Sciences* 6:256–259, 1991.

19. Rothwell NJ and Stock MJ. Insulin and thermogenesis. *International Journal of Obesity* 12:93–102, 1988.

20. Rothwell NJ and Stock MJ. Insulin and thermogenesis. *International Journal of Obesity* 12:93–102, 1988.

21. Kaats GR, Wise JA, Blum K, Morin RJ, Adelman JA, Craig J, and Croft HA. The short-term therapeutic efficacy of treating obesity with a plan of improved nutrition and moderate calorie restriction. *Current Therapeutic Research* 51:261–274, 1992.

22. Hass R. *Eat Smart, Think Smart.* HarperCollins: New York. 1994.

23. McCarty MF. The case for

supplemental chromium and a survey of clinical studies with chromium picolinate. *Journal of Applied Nutrition* 43:58–66, 1991.

Chapter 7
Chromium Picolinate and Diabetes

1. Ravina A and Slezack L. The clinical use of the trace element chromium (III) in the treatment of diabetes mellitus. *Harefuah* 125:142–148, 1993.
2. Lindemann MD, Harper AP, and Kornegay ET. Reproductive response in swine to the supplementation of chromium from chromium picolinate. *Journal of Animal Science* 72(Supp 1):66, 1994.
3. Lindemann MD, Wood CM, Harper AF, and Kornegay ET. Chromium picolinate addition to diets of growing-finishing pigs. *Journal of Animal Science* 71(Suppl 1):14, 1993.
4. Lindemann MD, Wood CM, Harper AF, Kornegay ET, and Anderson RA. Dietary chromium picolinate additions improve gain: feed and carcass characteristics in growing-finishing pigs and increase litter size in reproducing sows. *Journal of Animal Science* 73:457–465, 1995.
5. Kitchalong L, Fernandez JM, Bunting LD, Chapa AM, Sticker LS, Amoikon EK, Ward TL, Bidner TD, and Southern LL. Chromium picolinate supplementation in lamb rations. Effects on performance, nitrogen balance, endocrine and metabolic parameters. *Journal of Animal Science* 71(Supp 1):291, 1993.
6. Steele NC, Richards MP, and Rosebrough RW. Effect of dietary chromium and protein status on hepatic insulin binding characteristics of swine. *Journal of Animal Science* 55(Suppl 1):300, 1982.
7. Ward TL, Berrio LF, Southern LL, Fernandez JM, and Thompson DL Jr. In-vivo and in-vitro evaluation of chromium tripicolinate on insulin binding in pig liver cell plasma membranes. *FASEB Journal* 8(4):A194, 1994.
8. Evans GW and Bowman TD. Chromum picolinate increases membrane fluidity and rate of insulin internalization. *Journal of Inorganic Biochemistry* 46:243–250, 1992.
9. Evans GW and Pouchnik DJ. Composition and biological activity of chromium-pyridine carboxylate complexes. *Journal of Inorganic Biochemistry* 49:177–187, 1993.
10. Mertz W. Chromium occurrence and function in biological systems. *Physiological Reviews* 49:163–203, 1969.

11. Ravina A and Slezack L. The clinical use of the trace element chromium (III) in the treatment of diabetes mellitus. *Harefuah* 125:142–148, 1993.

12. Mertz W. Chromium occurrence and function in biological systems. *Physiological Reviews* 49:163–203, 1969.

13. Anderson RA. Chromium metabolism and its role in disease processes in man. *Clinical Physiology and Biochemistry* 4:31–41, 1986.

14. Freund H, Atamian S, and Fischer JE. Chromium deficiency during total parenteral nutrition. *Journal of the American Medical Association* 214:496–498, 1979.

15. Jeejeebhoy KN, Chu RC, Marliss EB, Greenberg R, and Bruce-Robertson AS. Chromium deficiency, glucose intolerance and neuropathy reversed by chromium supplementation in a patient receiving long-term total parenteral nutrition. *American Journal of Clinical Nutrition* 30:531–538, 1977.

16. Offenbacher EG and Pi-Sunyer FX. Beneficial effects of chromium-rich yeast on glucose tolerance and blood lipids in elderly subjects. *Diabetes* 29:919–925, 1980.

17. Pi-Sunyer FX and Offenbacher EG. Chromium. *Nutrition Reviews' Present Knowledge in Nutrition* 5:571–586, 1984.

18. Rabinowitz MB, Gonick HC, Levin SR, and Davidson MB. Clinical trial of chromium and yeast supplements on carbohydrate and lipid metabolism in diabetic men. *Biological Trace Element Research* 5:449–466, 1983.

19. Anderson RA, Polansky MM, Bryden NA, and Canary JJ. Chromium supplementation of humans with hypoglycemia. *Federation Proceedings* 41:471, 1984.

20. Pfeiffer CC. *Mental and Elemental Nutrients—A Physician's Guide to Nutrition and Health Care.* Keats Publishing: New Canaan, CT. 1975.

21. Dobbins JP. *Basics About Alcoholism.* Philosanus: Pasadena, CA. 1990.

22. Lindemann MD, Harper AP, and Kornegay ET. Reproductive response in swine to the supplementation of chromium from chromium picolinate. *Journal of Animal Science* 72(Supp 1):66, 1994.

23. Lindemann MD, Wood CM, Harper AF, and Kornegay ET. Chromium picolinate addition to diets of growing-finishing pigs. *Journal of Animal Science* 71(Suppl 1):14, 1993.

24. Lindemann MD, Wood CM, Harper AF, Kornegay ET, and Anderson RA. Dietary chromium picolinate addi-

tions improve gain: feed and carcass characteristics in growing-finishing pigs and increase litter size in reproducing sows. *Journal of Animal Science* 73:457–465, 1995.

25. Kitchalong L, Fernandez JM, Bunting LD, Chapa AM, Sticker LS, Amoikon EK, Ward TL, Bidner TD, and Southern LL. Chromium picolinate supplementation in lamb rations. Effects on performance, nitrogen balance, endocrine and metabolic parameters. *Journal of Animal Science* 71(Supp 1):291, 1993.

26. Steele NC, Richards MP, and Rosebrough RW. Effect of dietary chromium and protein status on hepatic insulin binding characteristics of swine. *Journal of Animal Science* 55(Suppl 1):300, 1982.

27. Ward TL, Berrio LF, Southern LL, Fernandez JM, and Thompson DL Jr. In-vivo and in-vitro evaluation of chromium tripicolinate on insulin binding in pig liver cell plasma membranes. *FASEB Journal* 8(4):A194, 1994.

28. Evans GW and Meyer LK. Life span is increased in rats supplemented with a chromium-pyridine 2 carboxylate complex. *Advances in Scientific Research* 1:19–23, 1994.

29. Lindemann MD, Harper AP, and Kornegay ET. Reproductive response in swine to the

supplementation of chromium from chromium picolinate. *Journal of Animal Science* 72(Supp 1):66, 1994.

30. Lindemann MD, Wood CM, Harper AF, and Kornegay ET. Chromium picolinate addition to diets of growing-finishing pigs. *Journal of Animal Science* 71(Suppl 1):14, 1993.

31. Lindemann MD, Wood CM, Harper AF, Kornegay ET, and Anderson RA. Dietary chromium picolinate additions improve gain: feed and carcass characteristics in growing-finishing pigs and increase litter size in reproducing sows. *Journal of Animal Science* 73:457–465, 1995.

32. Jovanovic-Peterson L, Gutierrez M, and Peterson CM. Chromium supplementation for gestational diabetic women improves glucose tolerance and decreases hyperinsulinemia. *Journal of the American College of Nutrition* 14:530, 1995.

Chapter 8
The Role of Chromium Picolinate in Sports

1. Evans GW and Pouchnik DJ. Composition and biological activity of chromium-pyridine carboxylate complexes. *Journal of Inorganic Biochemistry* 49:177–187, 1993.

2. Evans GW. Effect of chro-

mium picolinate on insulin controlled parameters in humans. *International Journal of Biosocial and Medical Research* 11:163–180, 1989.

3. Hasten DL, Rome EP, Franks BD, and Hegsted M. Effects of chromium picolinate on beginning weight training students. *International Journal of Sport Nutrition* 2:343–350, 1992.

4. Colgan M. *Optimum Sports Nutrition*. Advanced Research Press: Ronkonkoma, NY. 1993.

5. Hass R. *Eat Smart, Think Smart*. HarperCollins: New York. 1994.

6. Bahadori B, Habersack S, Schneider H, Wascher TC, and Toplak H. Treatment with chromium picolinate improves lean body mass in patients following weight reduction. *International Journal of Obesity* 19(Suppl2):38, 1995.

7. Page TG, Southern LL, Ward TL, and Thompson DL Jr. Effect of chromium picolinate on growth and serum and carcass traits of growing-finishing pigs. *Journal of Animal Science* 71:656–662, 1993.

8. Hasten DL, Hegsted M, and Glickman-Weiss EL. Effects of chromium picolinate on the body composition of the rat. *FASEB Journal* 7:A77, 1993.

9. Evans GW. The effect of exercise and chromium picoli-nate on dehydroepiandrosterone (DHEA) in males. Bemidji State University, 1992.

10. Evans GW and Pouchnik DJ. Composition and biological activity of chromium-pyridine carboxylate complexes. *Journal of Inorganic Biochemistry* 49:177–187, 1993.

11. Lefavi RG, Anderson RA, Keith RE, Wilson GD, McMillan JL, and Stone MH. Efficacy of chromium supplementation in athletes: emphasis on anabolism. *International Journal of Sport Nutrition* 2:111–122, 1992.

12. Evans GW and Pouchnik DJ. Composition and biological activity of chromium-pyridine carboxylate complexes. *Journal of Inorganic Biochemistry* 49:177–187, 1993.

13. Clancy S, Clarkson PM, DeCheke M, Nosaka K, Cunningham J, Freedson PS, and Valentine B. Chromium supplementation in football players. *Medicine and Science in Sports and Exercise* 25:S194, 1993.

14. Hallmark MA, Reynolds TH, DeSouza CA, Dotson CO, Anderson RA, and Rogers MA. Effects of chromium supplementation and resistive training on muscle strength and lean body mass in untrained men. *Medicine and Science in Sports and Exercise* 25:S101, 1993.

Chapter 9
*Using Chromium Picolinate
to Slow the Aging Process*

1. Cerami A. Hypothesis: Glucose as a mediator of aging. *Journal of the American Geriatric Society* 33:626–634, 1985.
2. Masoro EJ, Katz MS, and McMahan CA. Evidence for the glycation hypothesis of aging from the food-restricted rodent model. *Journal of Gerontology: Biological Science* 44:B20–B22, 1989.
3. McCay C, Crowell M, and Maynard L. The effect of retarded growth upon the length of life and upon the ultimate size. *Journal of Nutrition* 10:63–79, 1935.
4. Masoro EJ, Yu BP, and Bertrand HA. Action of food restriction in delaying the aging process. *Proceedings of the National Academy of Sciences, USA* 79:4239–4241, 1982.
5. Evans GW. Effect of chromium picolinate on insulin controlled parameters in humans. *International Journal of Biosocial and Medical Research* 11:163–180, 1989.
6. Evans GW and Meyer LK. Life span is increased in rats supplemented with a chromium-pyridine 2 carboxylate complex. *Advances in Scientific Research* 1:19–23, 1994.
7. Kalimi M and Regelson W. *The Biologic Role of Dehydroepiandrosterone (DHEA).* Walter de Gruyter & Co.: Berlin. 1990.
8. Nestler JE, Clore JN, and Blackard WG. Dehydroepiandrosterone: the "missing link" between hyperinsulinemia and atherosclerosis? *FASEB Journal* 6:3073–3075, 1992.
9. Dilman VM. Age-associated elevation of hypothalamic threshold to feedback control and its role in development, aging, and disease. *Lancet* i:1211–1219, 1971.
10. Everitt AV. The hypothalamic-pituitary control of aging and age-related pathology. *Experimental Gerontology* 8:265–277, 1973.
11. McCarty MF. Homologous physiological effects of phenformin and chromium picolinate. *Medical Hypotheses* 41:316–324, 1993.
12. Meites J, Goya R, and Takahashi S. Why the neuroendocrine system is important in aging processes. *Experimental Gerontology* 22:1–15, 1987.
13. Pierpaoli W, Dall'Ara A, Pedrinis E, and Regelson W. The pineal control of aging. The effects of melatonin and pineal grafting on the survival of older mice. *Annals of the New York Academy of Science* 621:291–313, 1991.
14. Piantanelli L, Gentile S, Fat-

toretti P, and Viticchi C. Thymic regulation of brain cortex beta-adrenoceptors during development and aging. *Archives of Gerontology and Geriatrics* 4:179–185, 1985.

15. Gonzalez C, Montoya E, and Jolin T. Effect of streptozotocin diabetes on the hypothalamic-pituitary-thyroid axis in the rat. *Endocrinology* 107:2099–2103, 1980.

16. Jackson RA. Mechanisms of age-related glucose intolerance. *Diabetes Care* 13(Suppl 2):9–19, 1990.

17. Cardarelli NF. The role of a thymus-pineal axis in an immune mechanism of aging. *Journal of Theoretical Biology* 145:397–405, 1990.

18. Cardarelli NF. The role of a thymus-pineal axis in an immune mechanism of aging. *Journal of Theoretical Biology* 145:397–405, 1990.

19. Schroeder HA. The role of chromium in mammalian nutrition. *American Journal of Clinical Nutrition* 21:230–244, 1968.

20. Evans GW and Meyer LK. Life span is increased in rats supplemented with a chromium-pyridine 2 carboxylate complex. *Advances in Scientific Research* 1:19–23, 1994.

21. Schroeder HA, Nason AP, and Tipton IH. Chromium deficiency as a factor in atherosclerosis. *Journal of Chronic Diseases* 23:123–142, 1970.

22. Schroeder HA. Serum cholesterol and glucose levels in rats fed refined and less refined sugars and chromium. *Journal of Nutrition* 97:237–242, 1969.

23. Schroeder HA. The role of chromium in mammalian nutrition. *American Journal of Clinical Nutrition* 21:230–244, 1968.

24. Hunt JV, Smith CCT, and Wolff SP. Autoxidative glycosylation and possible involvement of peroxides and free radicals in LDL modification by glucose. *Diabetes* 39:1420–1424, 1990.

25. Rawn JD. *Biochemistry*. Neil Patterson Publishers: Burlington, NC. 1989.

26. Schauf CL, Moffett DF, and Moffett SB. *Human Physiology*. Times Mirror Mosby College Publishing: St. Louis, MO. 1990.

27. Evans GW and Press RI. Cholesterol and glucose lowering effect of chromium picolinate. *FASEB Journal* 3:A761, 1989.

28. Press RI, Geller J, and Evans GW. The effect of chromium picolinate on serum cholesterol and apolipoprotein fractions in human subjects. *Western Journal of Medicine* 152:41–45, 1990.

29. Gordon JB. An easy and in-

expensive way to lower cholesterol? *Western Journal of
Medicine* 154:352, 1991.

30. Liarn TF, Chen SY, Shiau SP,
Froman DP, and Hu CY.
Chromium picolinate re-
duces serum and egg cholesterol of laying hen. *Professional Animal Science*, March
1996 (at press).

31. Newman HAI, Leighton RF,
Lanese RR, and Freedland
NA. Serum chromium and
angiographically determined
coronary artery disease.
Clinical Chemistry 24:541–544,
1978.

32. Simonoff M, Labador Y,
Hamon C, Peers AM, and
Nicolas-Simonoff G. Low
plasma chromium in pa-
tients with coronary artery
and heart diseases. *Biological
Trace Element Research* 6:431–
439, 1984.

33. Nestler JE, Clore JN, and
Blackard WG. Dehydro-
epiandrosterone: the "missing link" between hyperinsulinemia and atheros-
clerosis? *FASEB Journal*
6:3073–3075, 1992.

34. Verhaeghe J and Bouillon R.
Actions of insulin and the
IGFs on bone. *News In Physi-
ological Sciences* 9:20–22, 1994.

35. Evans GW, Swenson G, and
Walters K. Chromium picolinate decreases calcium excretion and increases dehy-
droepiandrosterone (DHEA)
in post menopausal women.
FASEB Journal 9:525, 1995.

Chapter 10
Chromium Picolinate for Life

1. Anderson RA, Bryden NA,
and Polansky MM. Dietary
chromium intake. Freely chosen diets, institutional diets
and individual foods. *Biological Trace Element Research*
32:117–121, 1992.

2. Kaats GR, Blum K, Fisher JA,
and Adelman JA. Effects of
chromium picolinate supplementation on body composition: a randomized double
blind placebo controlled
study. *Current Therapeutic Research*, February 1996 (at
press).

3. Kaats GR, Wise JA, Blum K,
Morin RJ, Adelman JA, Craig
J, and Croft HA. The short-
term therapeutic efficacy of
treating obesity with a plan
of improved nutrition and
moderate calorie restriction.
Current Therapeutic Research
51:261–274, 1992.

4. Ballor DL and Keesey RE. A
meta-analysis of the factors
affecting exercise-induced
changes in body mass, fat
mass and fat-free mass in
males and females. *International Journal of Obesity*
15:717–726, 1991.

5. Kaats GR, Wise JA, Blum K,
Morin RJ, Adelman JA, Craig

J, and Croft HA. The short-term therapeutic efficacy of treating obesity with a plan of improved nutrition and moderate calorie restriction. *Current Therapeutic Research* 51:261–274, 1992.

6. Anderson, JW. Nutrition management of diabetes mellitus. In Shils ME and Young VR (eds): *Modern Nutrition in Health and Disease.* Seventh edition. Lea and Febiger: Philadelphia, 1988.

7. Guise CC. Evaluation of chromium picolinate in the Ames/*Salmonella* mutagenesis assay. Biodevelopment Laboratories Reference Number 30101, 26 July 1995.

8. Anderson RA, Bryden NA, and Polansky MM. Lack of toxicity of chromium chloride and chromium picolinate. Paper to be presented at the Society of Toxicology, annual meeting, 1996.

9. Nakajima K. Japan Immuno Research Laboratories, letter to author, 15 September 1991.

10. Evans GW. Chromium picolinate is an efficacious and safe supplement. *International Journal of Sports Nutrition* 3:117–119, 1993.

11. Lindemann MD, Wood CM, Harper AF, Kornegay ET, and Anderson RA. Dietary chromium picolinate additions improve gain: feed and carcass characteristics in growing-finishing pigs and increase litter size in reproducing sows. *Journal of Animal Science* 73:457–465, 1995.

12. Evans GW and Meyer LK. Life span is increased in rats supplemented with a chromium-pyridine 2 carboxylate complex. *Advances in Scientific Research* 1:19–23, 1994.

13. Stearns DM, Wise JP, Patierno SR, and Wetterhahn KE. The dietary supplement chromium picolinate induces chromosome damage in Chinese hamster ovary cells. *FASEB Journal* 9:A451, 1995.

14. Sterns DM, Wise JP, Patierno SR, and Wetterhahn KE. Chromium (III) picolinate produces chromosome damage in Chinese hamster ovary cells. *FASEB Journal* 9:1643–1648, 1995.

15. Anderson RA, Bryden NA, and Polansky MM. Dietary chromium intake. Freely chosen diets, institutional diets and individual foods. *Biological Trace Element Research* 32:117–121, 1992.

16. Evans GW and Meyer LK. Life span is increased in rats supplemented with a chromium-pyridine 2 carboxylate complex. *Advances in Scientific Research* 1:19–23, 1994.

17. Anderson RA, Bryden NA, and Polansky MM. Form of chromium affects tissue

chromium concentrations. *FASEB Journal* 7:A204, 1993.

18. Evans GW and Pouchnik DJ. Composition and biological activity of chromium-pyridine carboxylate complexes. *Journal of Inorganic Biochemistry* 49:177–187, 1993.

GLOSSARY

Italicized terms are defined elsewhere in the Glossary.

Amino acid. One of the basic units from which proteins are made.

Anabolic steroid. A *hormone* that promotes muscular growth.

Atherosclerosis. The buildup of *cholesterol* on blood vessel walls. It contributes to heart attacks and strokes.

Basal metabolic rate (BMR). The unconcious processes, such as breathing, that keep the body alive. The BMR uses most of the *calories* consumed in a day.

Beta cells. The cells in the pancreas that produce *insulin*.

Bioavailability. The ability of the body to use a nutrient in a certain form. Chromium picolinate is a bioavailable form of *chromium*.

Biological impedance. A method of determining *lean body mass*, done by passing a weak electrical current through the skin.

Body composition. The amount of fat contained by the body versus the amount of *lean body mass*.

Calorie. The basic unit used to measure the energy content of food. A thousand calories equals one kilocalorie, or "Calorie." This book refers to "Calories" as "calories."

Carnitine. An *amino acid* that is needed before fat can be used for energy within the body. In at least one study, it helped improve the effectiveness of chromium picolinate.

Cholesterol. A fatty substance used by the body to build cell walls and create various necessary chemicals. Since cholesterol cannot dissolve in water, it must be combined with a *lipoprotein* to create either *HDL cholesterol* or *LDL cholesterol* before it can circulate in the bloodstream.

Chro Cal. A number that represents the amount of *chromium* compared to the number of *calories* in a given amount of food. It allows researchers to compare the amount of chromium used in animal studies with that used in human studies.

Chromium. A mineral that is needed in trace amounts to help *insulin* control blood sugar and fat. Since chromium by itself is not readily absorbed by the body, it must be combined with *picolinate* to be effective.

Cofactor. A substance that allows a protein to properly perform its function. *Chromium* is the cofactor for *insulin*.

Control. A test subject—either human or animal—that does not receive the substance being tested.

Cross-over. A medical study in which the participants take either the test substance pills or the *placebo* pills for a period of time and then switch to the other set of pills.

DHEA (dehydroepiandrosterone). A *hormone* that serves several functions in the body. It can be converted into *testosterone* in order to build muscles. It can also be converted into *estrogen* in order to prevent the loss of calcium through the urine.

Diabetes. The presence of *glucose* in the urine, a sign that the body can no longer control its use of glucose. *IDDM* and *NIDDM* are the two forms of diabetes related to *insulin*.

Double-blind. A medical study in which all the pills, both the test substance and the *placebo*, are in coded bottles, and neither the researchers nor the participants know which is which until the end of the test.

Endocrine gland. A gland that secretes a *hormone* such as *insulin*.

Estrogen. The primary female *hormone*, which serves several functions in the body. By preventing the loss of calcium through the urine, it can prevent *osteoporosis*.

Gestational diabetes. The high blood-*glucose* levels experienced by some women during pregnancy. It can lead to permanent *diabetes*.

Glucose. The simple sugar used by the body as its basic fuel. Also called "blood sugar."

Glucose intolerance. The body's inability to remove *glucose* from the bloodstream in a timely manner.

Glucose tolerance factor (GTF). The name originally given to the unknown material in yeast that corrected *glucose intolerance*. Chromium picolinate is probably the GTF.

Glucose tolerance test. A blood test that shows how quickly *glucose* is cleared from the bloodstream. It is used to detect *diabetes*.

Glucose transporter. A protein tunnel that carries *glucose* from the surface of a cell to the interior. It is activated when *insulin* attaches to a *receptor* on the cell membrane.

Glycation. The combination of *glucose* and various chemicals in the body, such as hemoglobin. Glycation occurs when there is too much glucose in the bloodstream.

Glycogen. A long chain of *glucose* molecules. It is the form in which glucose is stored by cells.

HDL (high-density lipoprotein) cholesterol. The "good" *cholesterol* that helps remove fat from the circulatory system.

Hormone. An *endocrine gland*-produced chemical that travels through the bloodstream instead of through a duct.

Hydrostatic weighing. A method of determining *lean body mass*, done by weighing the body in a special underwater chair.

Hyperglycemia. The condition that occurs when there is too much *glucose* in the bloodstream.

Hyperinsulinemia. The condition that occurs when there is too much *insulin* in the bloodstream.

Hypoglycemic shock. The condition that occurs when blood-*glucose* levels drop to dangerously low levels. It requires immediate medical attention.

Hypothalamus. An area of the brain that serves as one of the body's command centers. It contains the *satiety center*.

IDDM (insulin-dependent diabetes mellitus). The type of *diabetes* that results from the destruction of the body's *beta cells*. It is also called juvenile or Type 1 diabetes.

Insulin. The *hormone* that regulates the body's use of *glucose*, fat, and amino acids.

Insulin impotence. The inability of *insulin* to perform its function of controlling blood sugar and fat, linked to a lack of *chromium* in the diet.

LDL (low-density lipoprotein) cholesterol. The "bad" *cholesterol* that can gather on the walls of blood vessels, resulting in *atherosclerosis*.

Lean body mass. The amount of muscle, water, and bone the body contains, as compared with its fat content.

Lipoprotein. Proteins that carry fat and cholesterol through the bloodstream. See *HDL cholesterol* and *LDL cholesterol*.

Niacin. A vitamin (B3) that promotes proper functioning of the nervous, digestive, and circulatory systems. It has been shown in at least one study to improve the effectiveness of chromium picolinate in lowering cholesterol.

NIDDM (non-insulin-dependent diabetes mellitus). The type of *diabetes* that results from *insulin impotence*. It is also called adult-onset or Type 2 diabetes.

Obesity. The presence of excess body fat.

Osteoporosis. The loss of calcium from the bones, causing them to become brittle and easily cracked. It most often occurs in postmenopausal women.

Picolinate. An amino acid that serves as a chelator for *chromium* and other metal ions. It neutralizes chromium's electric charge, allowing the mineral to be absorbed by the body.

Pineal. A gland that controls the actions of the *hypothalamus* through the secretion of melatonin.

Placebo. An inert pill given to human study participants. The participants who receive the placebo act as test *controls*.

Reactive hypoglycemia. The temporary form of low blood sugar that results from the body's ineffective response to the presence of *glucose* in the bloodstream. It develops when blood-sugar levels become very high and then suddenly drop below normal.

Receptor. The protein "dock" on the cell membrane to which *insulin* attaches itself. This stimulates the activity of *glucose transporters*.

Satiety center. A group of cells in the *hypothalamus* that monitors the body's fuel supply.

Skin-fold test. A method of determining *lean body mass*, done by measuring the thickness of the skin on various parts of the body.

Testosterone. The most primary male *hormone*, which has several functions in the body. One of those functions is the building of muscle.

Thermogenesis. The production of heat by the body by the burning of excess glucose or fat.

Thymus. A gland that sets up the body's immune system in childhood and teaches certain kinds of white blood cells how to defend the body against attack.

Triglyceride. A chemical called glycerol attached to three strings of fat. It is the form in which fat cells store fat.

Tryptophan. The *amino acid* from which both *picolinate* and *niacin* are made.

INDEX

A

Acrodermatitis enteropathica (AE), 55, 57–58
Activation hormone. *See* Androgens.
Adult-onset diabetes. *See* Diabetes, non-insulin-dependent.
AE. *See* Acrodermatitis enteropathica.
Aging
   process of, 113–116
   program to slow, 122–123
Ames, Dr. Bruce, 132
Ames test, 132–133
Amino acids, 12, 19, 23, 53, 56, 66, 77, 103, 122. *See also* Carnitine; Tryptophan.
Anabolic hormones, 12

Anabolic steroids, 3
   synthetic, 18, 99, 100, 102–105
   See also DHEA.
Anderson, Dr. Richard, 92, 133
Androgens, 100, 103, 104, 105, 130
Asimov, Isaac, 67
Atherosclerosis, 8, 130
Athletic performance, chromium picolinate and, 109–110
Autonomic nerves, 115

B

Bahadori, Dr. B., 25
Banting, Dr. Frederick, 44
Basal metabolic rate (BMR), 82–83, 127–128

Healthy Habits
are easy to come by—
IF YOU KNOW WHERE TO LOOK!

Get the latest information on:
- **better health • diet & weight loss**
- **the latest nutritional supplements**
- **herbal healing • homeopathy and more**

COMPLETE AND RETURN THIS CARD RIGHT AWAY!

Where did you purchase this book?

❏ bookstore ❏ health food store ❏ pharmacy
❏ supermarket ❏ other (please specify)_____

Name_____

Street Address_____

City_____State_____Zip_____

RECEIVE A FREE
COPY OF
AVERY'S HEALTH
CATALOG

GIVE ONE TO A FRIEND ...

Healthy Habits
are easy to come by—
IF YOU KNOW WHERE TO LOOK!

Get the latest information on:
- **better health • diet & weight loss**
- **the latest nutritional supplements**
- **herbal healing • homeopathy and more**

COMPLETE AND RETURN THIS CARD RIGHT AWAY!

Where did you purchase this book?

❏ bookstore ❏ health food store ❏ pharmacy
❏ supermarket ❏ other (please specify)_____

Name_____

Street Address_____

City_____State_____Zip_____

RECEIVE A FREE
COPY OF
AVERY'S HEALTH
CATALOG

Avery Publishing Group
120 Old Broadway
Garden City Park, NY 11040

Avery Publishing Group
120 Old Broadway
Garden City Park, NY 11040